Springer Optimization and Its Applications 64

Series Editor:
Panos M. Pardalos

Subseries:
Nonconvex Optimization and Its Applications

T0137740

For further volumes:
http://www.springer.com/series/7393

Vladimir Shikhman

Topological Aspects of Nonsmooth Optimization

 Springer

Vladimir Shikhman
RWTH Aachen University
Mathematics, Section C
Templergraben 55
52056 Aachen, Germany
shikhman@mathc.rwth-aachen.de

ISSN 1931-6828
ISBN 978-1-4939-0069-5   ISBN 978-1-4614-1897-9 (eBook)
DOI 10.1007/978-1-4614-1897-9
Springer New York Dordrecht Heidelberg London

Mathematics Subject Classification (2010): 90C30, 90C31, 90C33, 90C34, 90C26, 57R45, 58K05, 49J52

Printed on acid-free paper

Springer is part of Springer Science+Business Media (www.springer.com)

for my grandmother Ina Kapilevich

Preface

The main goal of our study is an attempt to understand and classify nonsmooth structures arising within the optimization setting,

$$P(f,F): \quad \min f(x) \text{ s.t. } x \in M[F],$$

where $f : \mathbb{R}^n \longrightarrow \mathbb{R}$ is a smooth real-valued objective function, $F : \mathbb{R}^n \longrightarrow \mathbb{R}^l$ is a smooth vector-valued function, and $M[F] \subset \mathbb{R}^n$ is a feasible set defined by F in some structured way. The nonsmoothness is given by the structure that fits the smooth function F to define the feasible set $M[F]$. The following optimization problems with particular types of nonsmoothness are considered (Chapters 2–5):

- mathematical programming problems with complementarity constraints,
- general semi-infinite programming problems,
- mathematical programming problems with vanishing constraints,
- bilevel optimization.

The basis of our study is the topological approach introduced in detail in Chapter 1. It encompasses the following questions:

(a) Under which conditions on F is $M[F]$ a Lipschitz manifold of an appropriate dimension?
(b) Under which conditions on F is $M[F]$ stable (i.e., $M[F]$ remains invariant up to a homomorphism w.r.t. smooth perturbations of F)?
(c) How does the homotopy type of lower-level set

$$M[f,F]^a := \{x \in M[F] \,|\, f(x) \le a\}$$

change (as $a \in \mathbb{R}$ varies)?

Questions (a) and (b) deal with topological invariants of $M[F]$ and, more precisely, its structure. They lead to suitable constraint qualifications. Topological changes of $M[f,F]^a$ give rise to defining stationary points and developing critical point theory for $P(f,F)$ in the sense of Morse. In so doing, we get new topologically relevant optimization notions in terms of derivatives of f and F. It is worth pointing out that the same topological questions provide different (analytical) optimization concepts when applied to the particular problems above. The difference between these analytically described optimization concepts is a key point in understanding and comparing different kinds of nonsmoothness.

In Chapter 6, we discuss the impact of the topological approach on nonsmooth analysis. Topologically regular points of a min-type nonsmooth mapping $F : \mathbb{R}^n \longrightarrow \mathbb{R}^l$ are introduced. The crucial property is that for a topologically regular value $y \in \mathbb{R}^l$ of F the nonempty set $F^{-1}(y)$ is an $(n-l)$-dimensional Lipschitz manifold. Corresponding nonsmooth versions of Sard's Theorem are given.

We point out that the topological approach in the optimization context was introduced by H. Th. Jongen in the early 1980s ([61], [62]). The introduction of topological issues turned out to be extremely fruitful for establishing an adequate optimization theory in the smooth setting ([63]). The present book sheds light on nonsmooth optimization from the topological point of view, continuing to exploit the ideas of H. Th. Jongen.

I would like to thank my teacher H. Th. Jongen for sharing with me his insights on optimization and steering my studies toward its topological nature. This book originated mainly from a collaboration with him. I also thank my other coauthors, D. Dorsch, F. Guerra-Vázquez, Jan-J. Rückmann, S. Steffensen, and O. Stein, for fruitful collaborations. I am very grateful to H. Günzel, A. Ioffe, D. Klatte, B. Kummer, B. Mordukhovich, Yu. Nesterov, and D. Pallaschke for many interesting and helpful discussions.

Aachen, April 2011   *Vladimir Shikhman*

Contents

Notation

Our notation is standard. The n-dimensional Euclidean space is denoted by $\mathbb{R}^n$ with the norm $\| \cdot \|$, its nonnegative orthant by $\mathbb{H}^n$, and its nonpositive orthant by $\mathbb{R}^n_-$. $\mathbb{R}_+ := \{x \in \mathbb{R} \mid x > 0\}$. For $\varepsilon > 0$ and $\bar{x} \in \mathbb{R}^n$, the set $B_\varepsilon(\bar{x})$ (or $B(\bar{x}, \varepsilon)$) stands for the open Euclidean ball in $\mathbb{R}^n$ with radius ε and center $\bar{x}$. A closed ball with radius $\varepsilon > 0$ and center $\bar{x} \in \mathbb{R}^n$ is denoted by $\bar{B}(\bar{x}, \varepsilon)$.

Given an arbitrary set $K \subset \mathbb{R}^n$, $\overline{K}$, $\text{int}(K)$, and ∂K denote the topological closure, interior, and boundary of K, respectively. $\text{span}(K)$, $\text{conv}(K)$ (or $\text{co}(K)$), and $\text{cone}(K)$ denote the set of all linear, convex, and nonnegative combinations of elements of K, respectively. CK denotes the complement of $K \subset \mathbb{R}^n$. By $\text{span}\{a_1, \dots, a_t\}$ we denote the vector space over $\mathbb{R}$ generated by the finite number of vectors $a_1, \dots, a_t \in \mathbb{R}^n$, and $\dim\{\text{span}\{a_1, \dots, a_t\}\}$ stands for its dimension. The polar of K is defined by $K^\circ := \{v \in \mathbb{R}^n \mid v^T w \leq 0 \text{ for all } w \in K\}$. The distance from $x \in \mathbb{R}^n$ to $K \subset \mathbb{R}^n$ is denoted by $\text{dist}(x, K) = \inf_{y \in K} \|x - y\|$ with the convention $\text{dist}(x, \emptyset) = \infty$.

$T : \mathbb{R}^n \rightrightarrows \mathbb{R}^k$ denotes a multivalued map defined on $\mathbb{R}^n$ with $T(x) \subset \mathbb{R}^k$, $x \in \mathbb{R}^n$. The graph of T is $\text{gph } T = \{(x, y) \in \mathbb{R}^n \times \mathbb{R}^k \mid y \in T(x)\}$, and the inverse of T is $T^{-1} : \mathbb{R}^k \rightrightarrows \mathbb{R}^n$, given by $T^{-1}(y) = \{x \in \mathbb{R}^n \mid y \in T(x)\}$.

Given a differentiable function $F : \mathbb{R}^n \longrightarrow \mathbb{R}^k$, DF denotes its $k \times n$ Jacobian matrix. Given a differentiable function $f : \mathbb{R}^n \longrightarrow \mathbb{R}$, Df denotes its gradient as a row vector, and $D^T f$ (or ∇f) stands for the transposed vector. Given a twice continuously differentiable function $f : \mathbb{R}^n \longrightarrow \mathbb{R}$, $D^2 f$ stands for its Hessian. $C^l(\mathbb{R}^n, \mathbb{R}^k)$ denotes the space of l-times continuously differentiable functions from $\mathbb{R}^n$ to $\mathbb{R}^k$. $C^\infty(\mathbb{R}^n, \mathbb{R}^k)$ denotes the space of smooth functions from $\mathbb{R}^n$ to $\mathbb{R}^k$. $C^l(\mathbb{R}^n)$ stands for $C^l(\mathbb{R}^n, \mathbb{R})$, and $C^\infty(\mathbb{R}^n)$ stands for $C^\infty(\mathbb{R}^n, \mathbb{R})$.

Chapter 1
Introduction

We state mathematical programming problems with complementarity constraints, general semi-infinite programming problems, mathematical programming problems with vanishing constraints and bilevel optimization. The topological approach for studying problems above is introduced. It encompasses the study of topological properties of corresponding feasible sets, as well as the critical point theory in the sense of Morse. Finally, we describe the application of the topological approach for standard nonlinear programming problems.

1.1 Nonsmooth optimization framework

We consider the nonsmooth optimization framework

$$P(f,F): \quad \min f(x) \text{ s.t. } x \in M[F], \tag{1.1}$$

where $f : \mathbb{R}^n \longrightarrow \mathbb{R}$ is a real-valued objective function, $F : \mathbb{R}^n \longrightarrow \mathbb{R}^l$ is a vector-valued function, and $M[F] \subset \mathbb{R}^n$ is a feasible set defined by F in some structured way.

Within this general framework, the nonsmoothness might be caused by

(a)   the objective function f,
(b)   the defining function F, or
(c)   the structure according to which F defines $M[F]$.

Here, we assume functions f, F to be sufficiently smooth, and we restrict our study to the nonsmoothness given by (c). Thus, we focus rather on the underlying nonsmooth structures that fit the smooth function F to define the feasible set $M[F]$. We give some examples of particular optimization problems of type (1.1) to illustrate possible nonsmooth structures.

Example 1 (MPCC). The mathematical programming problem with complementarity constraints (MPCC) is defined as

1

$$\text{MPCC:} \quad \min f(x) \text{ s.t. } x \in M[h,g,F_1,F_2]$$

with

$$
\begin{aligned}
M[h,g,F_1,F_2] := \{x \in \mathbb{R}^n \mid &F_{1,m}(x) \geq 0, F_{2,m}(x) \geq 0,\\
&F_{1,m}(x)F_{2,m}(x) = 0, \, m = 1,\ldots,k,\\
&h_i(x) = 0, \, i \in I, \, g_j(x) \geq 0, \, j \in J\},
\end{aligned}
$$

where $f, h_i, i \in I, g_j, j \in J, F_{1,i}, F_{2,i}, i = 1,\ldots,k$ are real-valued and smooth functions, $|I| \leq n, |J| < \infty$.

Here, the nonsmoothness comes into play due to the complementarity constraints:

$$F_{1,m}(x) \geq 0, F_{2,m}(x) \geq 0, \, F_{1,m}(x)F_{2,m}(x) = 0, \, m = 1,\ldots,k.$$

Indeed, the basic complementarity relation

$$u \geq 0, \, v \geq 0, \, u \cdot v = 0,$$

defines the boundary of the nonnegative orthant in $\mathbb{R}^2$.

Example 2 (GSIP). Generalized semi-infinite programming problems (GSIPs) have the form

$$\text{GSIP:} \quad \text{minimize } f(x) \text{ s.t. } x \in M$$

with

$$M := \{x \in \mathbb{R}^n \mid g_0(x,y) \geq 0 \text{ for all } y \in Y(x)\}$$

and

$$Y(x) := \{y \in \mathbb{R}^m \mid g_k(x,y) \leq 0, \, k = 1,\ldots,s\}.$$

All defining functions $f, g_k, k = 0,\ldots,s$, are assumed to be real-valued and smooth on their respective domains.

Note that testing feasibility for x means that $\inf_{y \in Y(x)} g_0(x,y) \geq 0$. The appearance of the optimal value function $\inf_{y \in Y(x)} g_0(x,y)$ causes nonsmoothness.

Example 3 (MPVC). We consider the mathematical programming problem with vanishing constraints (MPVC)

$$\text{MPVC:} \quad \min f(x) \text{ s.t. } x \in M[h,g,H,G]$$

with

$$
\begin{aligned}
M[h,g,H,G] := \{x \in \mathbb{R}^n \mid &H_m(x) \geq 0, H_m(x)G_m(x) \leq 0, \, m = 1,\ldots,k,\\
&h_i(x) = 0, \, i \in I, \, g_j(x) \geq 0, \, j \in J\},
\end{aligned}
$$

where $f, h_i, i \in I, g_j, j \in J, H_m, G_m, m = 1,\ldots,k$ are real-valued and smooth functions, $|I| \leq n, |J| < \infty$.

Here, the difficulty is due to the vanishing constraints:

$$H_m(x) \geq 0, H_m(x)G_m(x) \leq 0, m = 1, \ldots, k.$$

Note that for those x with $H_m(x) = 0$ the sign of $G_m(x)$ is not restricted.

Example 4 (Bilevel optimization). We consider bilevel optimization from the optimistic point of view

$$U: \min_{(x,y)} f(x,y) \quad \text{s.t.} \quad y \in \text{Argmin } L(x),$$

where

$$L(x): \min_{y} g(x,y) \quad \text{s.t.} \quad h_j(x,y) \geq 0, j \in J.$$

Above we have $x \in \mathbb{R}^n$, $y \in \mathbb{R}^m$, and the real-valued mappings $f, g, h_j, j \in J$ are smooth, $|J| < \infty$. Argmin $L(x)$ denotes the solution set of the optimization problem $L(x)$.

Here, the nonsmoothness comes from the fact that we deal with a parametric nonlinear programming problem $L(x)$ at the lower level. Moreover, to ensure feasibility for (x,y) at the upper level U, the problem $L(x)$ should be solved up to global optimality.

1.2 Topological approach

The main goal of our study is an attempt to understand and classify nonsmooth structures arising in (1.1) within the optimization setting. The basis of such a comparison is the topological approach. It encompasses two objects of study:

the feasible set $M[F]$

and

the lower-level sets $M[f, F]^a := \{x \in M[F] \mid f(x) \leq a\}, a \in \mathbb{R}.$

These objects are considered along the levels of study due to topology, optimization and stability issues as outlined in the following scheme (see Figure 1).

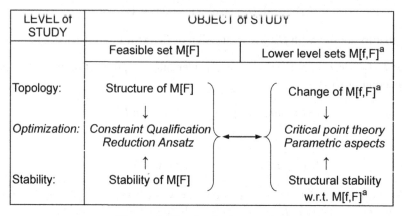

Figure 1 Topological approach

On the topology and stability levels we deal with topological invariants of $M[F]$ and $M[f,F]^a$, $a \in \mathbb{R}$. The questions mainly arise from here. They lead to establishment of an adequate theory on the optimization level. It is worth pointing out that the same topological questions provide different (analytical) optimization concepts when applied to particular problems (e.g., MPCC, GSIP, MPVC, and bilevel optimization). The difference between these analytically described optimization concepts is a key point in understanding and comparing different kinds of nonsmoothness. In what follows, we introduce the notions from the scheme in detail.

For the **structure of $M[F]$**, it is crucial to study under which conditions on F the feasible set is a **topological** or **Lipschitz manifold** (with boundary) of an appropriate dimension.

Definition 1 (Topological and Lipschitz manifolds [103]). A subset $\mathcal{M} \subseteq \mathbb{R}^n$ is called a topological (resp. Lipschitz) manifold (with boundary) of dimension $m \geq 0$ if for each $\bar{x} \in \mathcal{M}$ there exist open neighborhoods $U \subseteq \mathbb{R}^n$ of $\bar{x}$ and $V \subseteq \mathbb{R}^n$ of 0 and a homeomorphism $H : U \to V$ (resp. with H, H^{-1} being Lipschitz continuous) such that

(i) $H(\bar{x}) = 0$

and

(ii)   either in the first case

$$H(\mathcal{M} \cap U) = (\mathbb{R}^m \times \{0_{n-m}\}) \cap V$$

or in the second case

$$H(\mathcal{M} \cap U) = (\mathbb{H} \times \mathbb{R}^{m-1} \times \{0_{n-m}\}) \cap V$$

occur.

If for all $x \in \mathcal{M}$ the first case in (ii) holds, then $\mathcal{M}$ is called a topological (resp. Lipschitz) manifold of dimension m. In the second case, $\bar{x}$ is said to be a boundary point of $\mathcal{M}$ (see Figure 2).

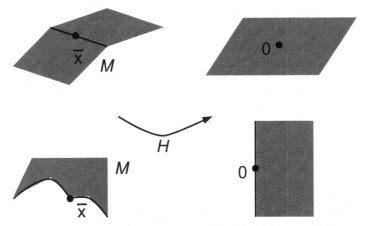

Figure 2 First and second cases for Lipschitz manifold

We shall use the **tools of nonsmooth and variational analysis** to tackle the question of $M[F]$ being a Lipschitz manifold. In particular, the application of nonsmooth versions of the **implicit function theorem** (see Section B.1) plays a major role.

Another issue for the structure of $M[F]$ is the **(topological) stability** of the feasible set under smooth perturbations of F (see Figure 3).

Definition 2 (Topological stability). The feasible set $M[F]$ from (1.1) is called (topologically) stable if there exists a C^1-neighborhood U of F in $C^1(\mathbb{R}^n, \mathbb{R}^l)$ (w.r.t. the strong or Whitney topology; see [42, 63], and Sections 1.3 and A.2 of the present volume) such that, for every $\widetilde{F} \in U$, the corresponding feasible set $M[\widetilde{F}]$ is homeomorphic with $M[F]$.

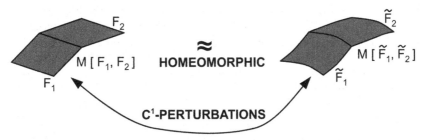

Figure 3 Topological stability

The stability of the feasible set is tightly connected with its Lipschitz manifold property. Addressing both of them will immediately lead us to suitable **constraint qualifications** for $M[F]$.

Actually, the list of topological invariants for $M[F]$ that is worth studying usually depends on particular problem realization. For example, having in mind GSIPs and bilevel optimization, an important issue for the description of the feasible set $M[F]$ becomes the so-called **reduction ansatz**. It deals with possibly infinite index sets that can be equivalently reduced to their finite subsets, at least at stationary points. Moreover, the feasible set in GSIPs need not be closed in general. This fact leads to the topological study of its closure instead. Next, the MPVC feasible set is not a Lipschitz manifold but a set glued together from manifold pieces of different dimensions along their strata.

Regarding the **behavior of the lower-level sets** $M[f, F]^a$, we study changes of their topological properties as $a \in \mathbb{R}$ varies. The smooth (un-) constrained case refers to the classical Morse theory and is well-known (see [63, 93]). We illustrate it in Figure 4.

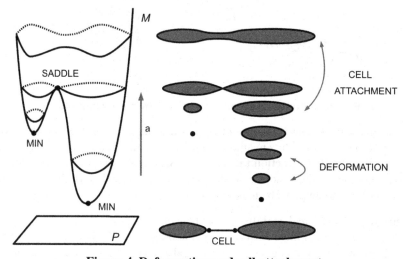

Figure 4 Deformation and cell attachment

Here, f is the height function from the plane P to the smooth manifold $M \subset \mathbb{R}^3$. Clearly, f has two local minima and one saddle point. We see that the topological changes of $M^a := \{x \in \mathbb{R}^2 \,|\, f(x) \leq a\}$, $a \in \mathbb{R}$ happen only when passing these three critical values. More precisely, new components of M^a are created passing local minima and, in addition, two components are attached together passing the saddle point. Note that the dimension of the cell attached corresponds to the number of negative eigenvalues of the Hessian of f.

Coming to the nonsmooth case, an adequate stationarity concept of **(topologically) stationary points** will be introduced. The analytical description of this concept depends certainly on a particular realization of (1.1). The definition of stationary points will be given in **dual terms** using Lagrange multipliers. Additionally, it will be shown that local minimizers are stationary points under some suitable constraint qualifications.

Within this context, two basic theorems from **Morse theory** (see [63, 93] and Section A.1) are crucial.

Theorem 1 (Deformation theorem). *If for* $a < b$ *the (compact) set* $M[f,F]_a^b :=$
$\{x \in M \mid a \le f(x) \le b\}$ *does not contain stationary points, then the set* $M[f,F]^a$ *is a strong deformation retract of* $M[f,F]^b$.

As a consequence, the homotopy types of the lower-level sets $M[f,F]^a$ and $M[f,F]^b$ are equal. This means that the connectedness structure of the lower-level sets does not change when passing from level a to level b. In particular, the number of connected (path) components remains invariant.

For the second result, a notion of a nondegenerate stationary point, along with its index, will be introduced. Note that a nondegenerate stationary point is a local minimizer if and only if its index vanishes.

Theorem 2 (Cell-attachment theorem). *If* $M[f,F]_a^b$ *contains exactly one nondegenerate stationary point, then* $M[f,F]^b$ *is homotopy-equivalent to* $M[f,F]^a$ *with a q-cell attached. Here, the dimension q is the so-called index of the nondegenerate stationary point.*

The latter two theorems on homotopy equivalence show that **Morse relations**, such as Morse inequalities (see [63]), are valid. Roughly speaking, Morse relations relate the existence of stationary points of various indices with the topology of the feasible set. In fact, the cell attachment of a k-dimensional cell can either generate a hole or cancel it (see Figure 5).

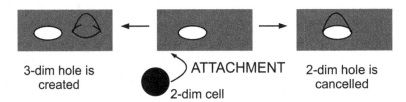

3-dim hole is created   ATTACHMENT   2-dim hole is cancelled

2-dim cell

Figure 5 Generation or cancellation of holes

A **global interpretation** of the deformation and cell-attachment theorems is the following. Suppose that the feasible set is compact and connected and that all stationary points are nondegenerate with pairwise different functional values. Then, passing a level corresponding to a local minimizer, a connected component of the lower-level set is created. Different components can only be connected by attaching 1-cells. This shows the existence of at least $(k-1)$ stationary points with index equal to 1, where k is the number of local minimizers; see also [26, 63]. This issue is closely related to the global aspects of optimization theory, in particular to the existence of $0-1-0$ and $0-n-0$ graphs. The latter connect local minimizers with stationary points having index equal to 1 and the former with local maximizers [63]. Finally, we refer the reader to [2, 6, 92] for the results with Morse theory for piecewise smooth functions.

The **structural stability w.r.t. lower-level sets** is defined via special equivalence relation on $P(f,F)$ as follows.

Definition 3 (Equivalence relation for optimization problems). Two optimization problems, $P(f,F)$ and $P\left(\widetilde{f},\widetilde{F}\right)$, are called equivalent if there exist continuous mappings $\varphi : \mathbb{R} \times \mathbb{R}^n \longrightarrow \mathbb{R}^n$, $\psi : \mathbb{R} \longrightarrow \mathbb{R}$ such that:

1. The mapping $\varphi(a,\cdot) : \mathbb{R}^n \longrightarrow \mathbb{R}^n$ is a homeomorphism for each $a \in \mathbb{R}$.
2. The mapping ψ is a homeomorphism and monotonically increasing.
3. For all $a \in R$, we have $\varphi\left(a, M[f,F]^a\right) = M[\widetilde{f},\widetilde{F}]^{\psi(a)}$.

The latter concept of equivalence was introduced in [31], and it was shown that it is indeed an equivalence relation.

Definition 4 (Structural stability). The optimization problem $P(f,F)$ is called structurally stable if there exists a C^2-neighborhood U of (f,F) in $C^2(\mathbb{R}^n, \mathbb{R}) \times C^2(\mathbb{R}^n, \mathbb{R}^l)$ (w.r.t. the strong or Whitney topology) such that, for every $(\widetilde{f},\widetilde{F}) \in U$, $P(\widetilde{f},\widetilde{F})$ is equivalent to $P(f,F)$.

The characterization of structural stability is related to both the stability of the feasible set (see Definition 2) and the **strong stability of stationary points** (in the sense of M. Kojima). The latter concept enlightens **parametric aspects** within the optimization context. In fact, it refers to existence, local uniqueness, and continuous dependence of stationary points w.r.t. local C^2-perturbations of the data functions (see Figure 6).

Definition 5 (Strong stability [83]). A stationary point $\bar{x} \in M[F]$ for $P(f,F)$ is called (C^2)-strongly stable if for some $r > 0$ and each $\varepsilon \in (0,r]$ there exists a $\delta = \delta(\varepsilon) > 0$ such that, whenever $\left(\widetilde{f},\widetilde{F}\right) \in C^2$ and $\left\| \left(f - \widetilde{f}, F - \widetilde{F}\right) \right\|_{B(\bar{x},r)}^{C^2} \leq \delta$, the ball $B(\bar{x},\varepsilon)$ contains a stationary point $\widetilde{x}$ for $P\left(\widetilde{f},\widetilde{F}\right)$ that is unique in $B(\bar{x},r)$.

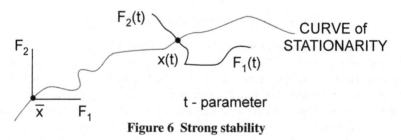

Figure 6 Strong stability

1.3 Genericity and stability issues

Our goal is to justify the assumptions made on the optimization level such as constraint qualifications, reduction ansatz, and nondegeneracy of stationary points. For that, genericity and stability issues w.r.t. the strong or Whitney topology on defining functions (f,F) come into play.

Let $C^k(\mathbb{R}^n, \mathbb{R})$, $k = 0, 1, \ldots$, denote the space of k-times continuously differentiable real-valued functions. Let $C^k(\mathbb{R}^n, \mathbb{R})$ be endowed with the strong (or Whitney) C^k-topology, denoted by C_s^k (see [42, 63] and Section A.2). The C_s^k-topology is generated by allowing perturbations of the functions and their derivatives up to k-th order, which are controlled by means of continuous positive functions. The product space of continuously differentiable functions will be topologized with the corresponding product topology. Note that the space of continuously differentiable functions endowed with the strong C_s^k-topology constitutes a Baire space. We say that a set is C_s^k-generic if it contains a countable intersection of C_s^k-open and C_s^k-dense subsets. Generic sets in a Baire space are dense as well.

Next, we explain a typical application of C_s^k-topology in the optimization context. Let $\mathfrak{A}$ be an assumption involving derivatives of (f, F) up to the k-th order in its formulation (e.g., constraint qualification, conditions in reduction ansatz, nondegeneracy, etc.) We are interested in the following type of result.

Theorem 3 (Assumption $\mathfrak{A}$ is generic and stable). *Let $\mathscr{A}$ denote the set of problem data $(f, F) \in C^k(\mathbb{R}^n, \mathbb{R}) \times C^k(\mathbb{R}^n, \mathbb{R}^l)$ such that the assumption $\mathfrak{A}$ is satisfied. Then, $\mathscr{A}$ is C_s^k-generic and C_s^k-open.*

Theorem 3 can be interpreted as follows. If Assumption $\mathfrak{A}$ does not hold for the concrete problem data (f, F), then we may find an arbitrary small perturbed problem $\left(\widetilde{f}, \widetilde{F}\right)$ with Assumption $\mathfrak{A}$ fulfilled. Moreover, if Assumption $\mathfrak{A}$ holds for (f, F), then it also holds for sufficiently small perturbations of (f, F). We point out that the proofs of such results are mainly based on **transversality theory**, in particular on Thom's transversality theorem (see [42, 63, 90] and Section B.2).

The justification of assumptions w.r.t. genericity and stability issues does not certainly exclude the study of so-called singular situations. The latter are characterized by the fact that a certain assumption is not fulfilled. One speaks also of problem data not in general position. This phenomenon is studied within the scope of **singularity theory** (see [3, 11, 27, 28]). It remains very challenging to apply ideas from singularity theory in the optimization context (see [4, 58, 54]). We will touch on this topic when studying bilevel optimization.

1.4 Nonlinear programming: smooth case

We illustrate the application of the topological approach in its entirety for the smooth case. For that, we consider the nonlinear programming (NLP) problem:

$$\text{NLP}(f, g, h): \quad \min f(x) \text{ s.t. } x \in M[h, g] \tag{1.2}$$

with

$$M[h, g] := \{x \in \mathbb{R}^n \mid h_i(x) = 0, i \in I, g_j(x) \geq 0, j \in J\},$$

where $h := (h_i, i \in I)^T \in C^2(\mathbb{R}^n, \mathbb{R}^{|I|})$, $g := (g_j, j \in J)^T \in C^2(\mathbb{R}^n, \mathbb{R}^{|J|})$, and $f \in C^2(\mathbb{R}^n, \mathbb{R})$ with $|I| \leq n$, $|J| < \infty$.

The optimization theory for the NLP is very well known (e.g., [63]). We briefly recapitulate how the main questions of the topological approach lead to optimization theory.

Stability and structure of the feasible set

The feasible set $M[h,g]$ from (1.1) is called **(topologically) stable** if there exists a C^1-neighborhood U of (h,g) in $C^1(\mathbb{R}^n, \mathbb{R}^{|I|+|J|})$ w.r.t. the strong or Whitney topology such that for every $(\widetilde{h}, \widetilde{g}) \in U$ the corresponding feasible set $M[\widetilde{h}, \widetilde{g}]$ is homeomorphic with $M[h,g]$. For the NLP, the topological stability of the feasible set corresponds to the well-known **Mangasarian-Fromovitz constraint qualification** (MFCQ). Recall that the MFCQ is said to hold at $\bar{x} \in M[h,g]$ if

(1) $D^T h_i(\bar{x})$, $i \in I$ are linearly independent

and

(2)   there exists a vector $\xi \in \mathbb{R}^n$ satisfying

$$D^T h_i(\bar{x})\xi = 0, i \in I,$$
$$D^T g_j(\bar{x})\xi > 0, j \in J_0(\bar{x}) := \{j \in J \,|\, g_j(\bar{x}) = 0\}.$$

Theorem 4 (Stability of the feasible set for NLP [32, 67]). *Suppose that $M[h,g]$ is compact. Then, the feasible set $M[h,g]$ is stable if and only if the MFCQ holds at every point $x \in M[h,g]$.*

The feasible set turns out to be a Lipschitz manifold (with boundary) under the MFCQ.

Theorem 5 (Structure of the feasible set for NLP; e.g., [77]). *Suppose that the MFCQ is satisfied at all points of the compact set $M[h,g]$. Then, the feasible set $M[h,g]$ is a Lipschitz manifold (with boundary) of dimension $n - |I|$, and the boundary $\partial M[h,g]$ equals $\{x \in \mathbb{R}^n \,|\, h_i(x) = 0, i \in I, \min_{j \in J} g_j(x) = 0\}$.*

We point out that the MFCQ is a generic property for the NLP and hence fairly typical.

Theorem 6 (MFCQ is generic for NLP; e.g., [63]). *Let $\mathscr{F}$ denote the subset of $C^1(\mathbb{R}^n)^{|I|+|J|}$ consisting of those problem data (h,g) for which the MFCQ holds at all points $x \in M[h,g]$. Then, $\mathscr{F}$ is C^1_s-open and C^1_s-dense.*

Critical point theory

We need some preliminary notions to formulate the main results of the critical point theory for the NLP.

The well-known **linear independence constraint qualification (LICQ) for the NLP** is said to hold at $\bar{x} \in M[h,g]$ if the set of vectors

$$\{D^T h_i(\bar{x}), i \in I, D^T g_j(\bar{x}), j \in J_0(\bar{x})\}$$

is linearly independent. A point $\bar{x} \in M[h,g]$ is called **Karusch-Kuhn-Tucker** (KKT) if there exist real numbers $\bar{\lambda}_i, i \in I, \bar{\mu}_j, j \in J_0(\bar{x})$ (Lagrange multipliers) such that

$$Df(\bar{x}) = \sum_{i \in I} \bar{\lambda}_i Dh_i(\bar{x}) + \sum_{j \in J_0(\bar{x})} \bar{\mu}_j Dg_j(\bar{x})$$

and

$$\bar{\mu}_j \geq 0 \text{ for all } j \in J_0(\bar{x}).$$

In the case where the LICQ holds at $\bar{x} \in M[h,g]$, the Lagrange multipliers are uniquely determined. A **KKT point** $\bar{x} \in M[h,g]$ with Lagrange multipliers as above is called **nondegenerate** if the following conditions are satisfied:

ND1:   The LICQ holds at $\bar{x}$.

ND2: $\bar{\mu}_j > 0$ for all $j \in J_0(\bar{x})$ (strict complementarity).

ND3: $D^2_{xx} L(\bar{x}, \bar{\lambda}, \bar{mu})\,|_{T_{\bar{x}} M[h,g]}$ is nonsingular (second-order nondegeneracy).

Here, the matrix $D^2 L$ stands for the Hessian of the Lagrange function L,

$$L(x, \bar{\lambda}, \bar{\mu}) := f(x) - \sum_{i \in I} \bar{\lambda}_i h_i(x) - \sum_{j \in J_0(\bar{x})} \bar{\mu}_j g_j(x),$$

and $T_{\bar{x}} M[h,g]$ denotes the tangent space of $M[h,g]$ at $\bar{x}$. Condition ND3 means that the matrix $V^T D^2 L(\bar{x}) V$ is nonsingular, where V is some matrix whose columns form a basis for the tangent space $T_{\bar{x}} M[h,g]$. Furthermore, let $\bar{x} \in M[h,g]$ be a nondegenerate KKT point with Lagrange multipliers as above. The number of negative/positive eigenvalues of $D^2 L(\bar{x})\,|_{T_{\bar{x}} M[h,g]}$ is called the **quadratic index (QI)/quadratic coindex (QCI)** of $\bar{x}$.

The following proposition uses the quadratic index for the characterization of a local minimizer.

Proposition 1 (Local minimizers and quadratic index; see [65]).

(i)   *Suppose that $\bar{x}$ is a local minimizer for the NLP and the LICQ holds at $\bar{x}$. Then, $\bar{x}$ is a KKT point.*

(ii) *Let $\bar{x}$ be a nondegenerate KKT point. Then, $\bar{x}$ is a local minimizer for the NLP if and only if its quadratic index is equal to zero.*

The next proposition concerning genericity results for the LICQ and nondegeneracy of KKT points is due to [63].

Theorem 7 (Genericity and stability for NLP; see [63]).

(i)   *Let $\mathscr{F}$ denote the subset of $C^1(\mathbb{R}^n)^{|I|+|J|}$ consisting of those (h,g) for which the LICQ holds at all points $x \in M[h,g]$. Then, $\mathscr{F}$ is C^1_s-open and -dense.*

(ii)   Let $\mathcal{D}$ denote the subset of $C^2(\mathbb{R}^n, \mathbb{R}) \times C^2(\mathbb{R}^n)^{|I|+|J|}$ consisting of those (f, h, g) for which each KKT point for the NLP with data functions (f, h, g) is nondegenerate. Then, $\mathcal{D}$ is C_s^2-open and -dense.

Finally, we state the main deformation and cell-attachment theorem for the critical point theory of the NLP. Recall that for $a, b \in \mathbb{R}$, $a < b$, the sets M^a and M_a^b are defined as

$$M^a := \{x \in M[h, g] \mid f(x) \le a\}$$

and

$$M_a^b := \{x \in M[h, g] \mid a \le f(x) \le b\}.$$

Theorem 8. Let M_a^b be compact, and suppose that the LICQ is satisfied at all points $x \in M_a^b$.

(a)   **(Deformation theorem)** If M_a^b does not contain any KKT point, then M^a is a strong deformation retract of M^b.

(b)   **(Cell-attachment theorem)** If M_a^b contains exactly one KKT point, say $\bar{x}$, and if $a < f(\bar{x}) < b$ and the quadratic index of $\bar{x}$ is equal to q, then M^b is homotopy-equivalent to M^a with a q-cell attached.

Parametric aspects

We recapitulate the characterization of the **structural stability w.r.t. lower-level sets** for the NLP (see [31, 78]). This concept is defined via a special equivalence relation on $P(f, h, g)$ (see Definitions 3 and 4) as follows. Two **optimization problems**, NLP(f, g, h) and NLP$\left(\widetilde{f}, \widetilde{h}, \widetilde{g}\right)$, are called **equivalent** if there exist continuous mappings $\varphi : \mathbb{R} \times \mathbb{R}^n \longrightarrow \mathbb{R}^n$, $\psi : \mathbb{R} \longrightarrow \mathbb{R}$ such that:

1.   The mapping $\varphi(a, \cdot) : \mathbb{R}^n \longrightarrow \mathbb{R}^n$ is a homeomorphism for each $a \in \mathbb{R}$.
2.   The mapping ψ is a homeomorphism and monotonically increasing.
3.   For all $a \in R$, we have $\varphi\left(a, M[f, h, g]^a\right) = M[\widetilde{f}, \widetilde{h}, \widetilde{g}]^{\psi(a)}$.

The optimization problem NLP(f, F) is called **structurally stable** if there exists a C^2-neighborhood U of (f, h, g) in $C^2(\mathbb{R}^n, \mathbb{R}) \times C^2(\mathbb{R}^n)^{|I|+|J|}$ (w.r.t. the strong or Whitney topology) such that, for every $(\widetilde{f}, \widetilde{h}, \widetilde{g}) \in U$, NLP$(\widetilde{f}, \widetilde{h}, \widetilde{g})$ is equivalent to NLP(f, h, g).

The characterization of structural stability rests on the notion of **strong stability of KKT points** (in the sense of M. Kojima; see Definition 5). Namely, a KKT point $\bar{x} \in M[h, g]$ for NLP(f, h, g) is called (C^2)-**strongly stable** if for some $r > 0$ and each $\varepsilon \in (0, r]$ there exists a $\delta = \delta(\varepsilon) > 0$ such that, whenever $\left(\widetilde{f}, \widetilde{h}, \widetilde{g}\right) \in C^2$ and $\left\| \left(f - \widetilde{f}, h - \widetilde{h}, g - \widetilde{g}\right) \right\|_{B(\bar{x}, r)}^{C^2} \le \delta$, the ball $B(\bar{x}, \varepsilon)$ contains a KKT point $\widetilde{x}$ for NLP$\left(\widetilde{f}, \widetilde{h}, \widetilde{g}\right)$ that is unique in $B(\bar{x}, r)$.

Theorem 9 (Strong stability of KKT points; see [35, 83]). *A KKT point $\bar{x} \in M[h,g]$ is strongly stable if and only if the following conditions are satisfied:*

(i)   *The MFCQ holds at $\bar{x}$.*
(ii)   *For all Lagrange multipliers (λ, μ) and all index sets J' with*

$$J_+(\mu) := \{j \in J \mid \mu_j > 0\} \subset J' \subset J_0(\bar{x}),$$

the matrix $D^2_{xx} L(\bar{x}, \lambda, \mu)\mid_{T^{J'}}$ is nonsingular and has the same number of negative eigenvalues $ind_-(\bar{x})$. Here,

$$T^{J'} := \{\xi \in \mathbb{R}^n \mid Dh_i(\bar{x})\xi = 0, \ i \in I, \ Dg_j(\bar{x})\xi = 0, \ i \in J'\}.$$

(iii) *If the LICQ fails to hold at $\bar{x}$, then $ind_-(\bar{x}) = 0$.*

Now we are ready to state the structural stability characterization for NLP. It is tightly related to both the stability of the feasible set and the strong stability of KKT points.

Theorem 10 (Structural stability for NLP; see [31, 78]). *Let the feasible set $M[h,g]$ corresponding to $NLP(f,h,g)$ be compact. Then, the nonlinear programming problem $NLP(f,h,g)$ is structurally stable if and only if the conditions S1, S2, S3 are satisfied:*

(S1)   *The MFCQ holds at all points $x \in M[h,g]$.*
(S2)   *Every KKT point is strongly stable.*
(S3)   *Different KKT points have different f-values.*

Chapter 2
Mathematical Programming Problems with Complementarity Constraints

We study mathematical programming problems with complementarity constraints (MPCC) from the topological point of view. The (topological) stability of the MPCC feasible set is addressed. Therefore, we introduce Mangasarian-Fromovitz condition (MFC) and its stronger version (SMFC). Under SMFC, the MPCC feasible set is shown to be a Lipschitz manifold. The links to other well-known constraint qualifications for MPCCs are elaborated. The critical point theory for MPCCs is presented. We also characterize the strong stability of C-stationary points for MPCC, dealing with parametric aspects for MPCCs.

2.1 Applications and examples

We consider the mathematical programming problem with complementarity constraints (MPCC)

$$\text{MPCC:} \quad \min f(x) \text{ s.t. } x \in M[h, g, F_1, F_2] \tag{2.1}$$

with

$$
\begin{aligned}
M[h, g, F_1, F_2] := \{x \in \mathbb{R}^n \mid\ & F_{1,m}(x) \geq 0, F_{2,m}(x) \geq 0, \\
& F_{1,m}(x)F_{2,m}(x) = 0,\ m = 1, \ldots, k, \\
& h_i(x) = 0,\ i \in I,\ g_j(x) \geq 0,\ j \in J\},
\end{aligned}
$$

where $h := (h_i, i \in I)^T \in C^2(\mathbb{R}^n, \mathbb{R}^{|I|})$, $g := (g_j, j \in J)^T \in C^2(\mathbb{R}^n, \mathbb{R}^{|J|})$, $F_1 := (F_{1,i}, i = 1, \ldots, k)^T, F_2 := (F_{2,i}, i = 1, \ldots, k)^T \in C^2(\mathbb{R}^n, \mathbb{R}^k), f \in C^2(\mathbb{R}^n, \mathbb{R}), k + |I| \leq n, |J| < \infty$. For simplicity, we write M for $M[h, g, F_1, F_2]$ if no confusion is possible.
For $m = 1, \ldots, k$, the constraint

$$F_{1,m}(x) \geq 0, F_{2,m}(x) \geq 0, F_{1,m}(x)F_{2,m}(x) = 0$$

is called a complementarity constraint. Note that it can be equivalently written as $\min\{F_{1,m}(x), F_{2,m}(x)\} = 0$.

The MPCC is a special case of the so-called mathematical programming problem with equilibrium constraints (MPEC) (see [88]). In what follows, we show that MPCCs appear quite naturally in bilevel optimization (via Karush-Kuhn-Tucker or Fritz John conditions at the lower level) and when solving nonlinear complementarity problems. Moreover, complementarity constraints arise in the context of variational inequalities. For other applications, we refer the reader to [23, 88, 97].

Bilevel optimization with convexity at the lower level

We model the bilevel optimization problem in the so-called optimistic formulation. To this aim, assume that the follower solves the parametric optimization problem (lower-level problem L)

$$L(x): \quad \min_y g(x,y) \quad \text{s.t.} \quad h_j(x,y) \geq 0, \, j \in J$$

and that the leader's optimization problem (upper-level problem U) is

$$U: \quad \min_{(x,y)} f(x,y) \quad \text{s.t.} \quad y \in \text{Argmin } L(x).$$

Above we have $x \in \mathbb{R}^n$, $y \in \mathbb{R}^m$, and the real-valued mappings $f, g, h_j, j \in J$ belong to $C^2(\mathbb{R}^n \times \mathbb{R}^m)$, $|J| < \infty$. Argmin $L(x)$ denotes the solution set of the optimization problem $L(x)$. For simplicity, additional (in)equality constraints in defining U are omitted.

We assume convexity at the lower level $L(\cdot)$; that is, for all $x \in \mathbb{R}^n$, let the functions $g(x, \cdot)$, $-h_j(x, \cdot)$, $j \in J$ be convex. For example, the Slater constraint qualification (CQ) hold for $L(\cdot)$. Then, it is well-known that $y \in \text{Argmin } L(x)$ if and only if there exist Lagrange multipliers $\mu_j \in \mathbb{R}$, $j \in J$ such that

$$D_y g(x,y) = \sum_{j \in J} \mu_j D_y h_j(x,y), \, \mu_j \geq 0, \, h_j \geq 0, \, \mu_j h_j(x,y) = 0. \qquad (2.2)$$

Hence, we can write the corresponding MPCC:

$$U\text{-KKT}: \quad \min_{(y,\mu) \in \mathbb{R}^m \times \mathbb{R}^{|J|}} g(x,y) \quad \text{s.t.}$$

$$D_y g(x,y) = \sum_{j \in J} \mu_j D_y h_j(x,y), \, \mu_j \geq 0, h_j \geq 0, \mu_j h_j(x,y) = 0.$$

Here, the complementarity constraints are $\mu_j \geq 0$, $h_j \geq 0$, $\mu_j h_j(x,y) = 0$.

The links between U and U-KKT were elaborated in [16]. It turns out that global solutions of U and U-KKT coincide. But, due to the possible non-uniqueness of Lagrange multipliers in (2.2), local solutions of U and U-KKT may differ.

Note that it is very restrictive to assume the Slater CQ in $L(x)$ for all $x \in \mathbb{R}^n$. Hence, one may try to assume the Slater CQ only at the point of interest $\bar{x}$. However, in that case even global solutions of U and U-KKT may differ (as shown in [16]).

Without assuming the Slater CQ, we arrive at the MPCC-relaxation of U:

$$U\text{-John}: \quad \min_{(y,\delta,\mu) \in \mathbb{R}^m \times \mathbb{R} \times \mathbb{R}^{|J|}} g(x,y) \quad \text{s.t.}$$

$$\delta D_y g(x,y) = \sum_{j \in J} \mu_j D_y h_j(x,y), \tag{2.3}$$

$$\mu_j D_y h_j(x,y), \ \mu_j \geq 0, \ h_j \geq 0, \ \mu_j h_j(x,y) = 0, \ \delta \geq 0.$$

Here, we use the fact that $y \in \text{Argmin } L(x)$ fulfills the Fritz John condition. In fact, generically one cannot exclude the violation of the LICQ or even MFCQ at the lower level. Thus, the case of vanishing δ in (2.3) cannot be omitted (see Chapter 5 for details).

Solving nonlinear complementarity problems

We consider a nonlinear complementarity problem (NCP) of finding $x \in \mathbb{R}^n$ such that

$$x \geq 0, \ F(x) \geq 0, \ x^T F(x) = 0,$$

where $F : \mathbb{R}^n \longrightarrow \mathbb{R}^n$ is continuously differentiable. Such problems appear in many applications, such as equilibria models of economics, contact and structural mechanics problems, and obstacle problems (see also [98]).

Setting $H(x) := \min\{x, F(x)\}$ componentwise, we obtain a residual optimization problem

$$\text{RES}: \quad \min_x \vartheta(x) := \frac{1}{2} H(x)^T H(x) \text{ s.t. } x \geq 0.$$

Obviously, if $\bar{x}$ is a solution of an NCP, then $\bar{x}$ is a solution of RES with $\vartheta(\bar{x}) = 0$. Moreover, ϑ is nonnegative and vanishes exactly at solutions of the NCP.

With $y := x - \min\{x, F(x)\}$, it is easy to see that RES can be equivalently written as an MPCC:

$$\text{RES-MPCC}: \quad \min_{(x,y)} \frac{1}{2}(x-y)^T(x-y) \text{ s.t.}$$

$$x \geq 0, \ y \geq 0, \ F(x) - x - y \geq 0, \ y^T(F(x) - x - y) = 0.$$

This problem is used to solve an NCP numerically (see [88]).

Variational inequalities setting

Let $K \subset \mathbb{R}^n$ and $F : K \longrightarrow \mathbb{R}^n$ be given. The variational inequality $VI(K,F)$ is the following problem:

$VI(K,F)$:   Find $x \in \mathbb{R}^n$ such that $(y-x)^T F(x) \geq 0$ for all $y \in K$

Clearly, $\bar{x}$ is a solution of $VI(K,F)$ if and only if

$$0 \in F(\bar{x}) + N(\bar{x}, K), \tag{2.4}$$

where $N(\bar{x}, K)$ is a normal cone of K at $\bar{x}$:

$$N(\bar{x}, K) := \left\{ d \in \mathbb{R}^n \,|\, d^T(\bar{x} - y) \leq 0 \text{ for all } y \in K \right\}.$$

Equation (2.4) can be seen as a generalization of the first-order optimality conditions to minimize a differentiable function $f : \mathbb{R}^n \longrightarrow R$ on a convex set K.

Furthermore, if K is a cone, we may link variational inequalities with so-called complementarity problems:

$CP(K,F)$:   Find $x \in \mathbb{R}^n$ such that $x \in K$, $F(x) \in K^\star$, $X^T F(x) = 0$,

where $K^\star := \left\{ d \in \mathbb{R}^n \,|\, v^T d \geq 0 \text{ for all } v \in K \right\}$ is a dual cone of K.

It can be shown (see, e.g., [22]) that, in the case of K being a cone, solutions of $VI(K,F)$ and $CP(K,F)$ coincide. Moreover, let

$$K := \{ x \in \mathbb{R}^n \,|\, Ax \leq b, \, Cx = d \}$$

with matrices $A \in \mathbb{R}^{m \times n}$, $B \in \mathbb{R}^{l \times n}$ and vectors $b \in \mathbb{R}^m$, $d \in \mathbb{R}^l$. Then, $\bar{x}$ solves $VI(K,F)$ if and only if there exist $\lambda \in \mathbb{R}^m$, $\mu \in R^l$ such that

$$F(x) + A^T \lambda + C^T \mu = 0, \, C - dx = 0,$$

$$\lambda \geq 0, \, b - Ax \geq 0, \, \lambda^T(b - Ax) = 0.$$

The latter system exhibits complementarity constraints and hence fits in the context of an MPCC.

Note that, setting $K := \mathbb{H}^n$ in $CP(K,F)$, we obtain the usual nonlinear complementarity problem.

2.2 Stability and structure of the feasible set

In this section, we concentrate only on the substantial new case of complementarity constraints. Hence, we omit smooth equality and inequality constraints and consider the mathematical programming problem with complementarity constraints (MPCC):

MPCC: $\min f(x)$ s.t. $x \in M[F_1, F_2]$ (2.5)

with

$$M[F_1, F_2] := \{ x \in \mathbb{R}^n \,|\, F_1(x) \geq 0, F_2(x) \geq 0, F_1(x)^T F_2(x) = 0 \},$$

where $F_1 := (F_{1,i}, i = 1, \ldots, k)^T, F_2 := (F_{2,i}, i = 1, \ldots, k)^T \in C^1(\mathbb{R}^n, \mathbb{R}^k)$, and $f \in C^1(\mathbb{R}^n, \mathbb{R})$ with $k \leq n$.

Note that $M[F_1, F_2]$ can be written as

$$M[F_1, F_2] = \{x \in \mathbb{R}^n \mid \min\{F_{1,i}(x), F_{2,i}(x)\} = 0, i = 1, \ldots, k\}.$$

Here we deal with the local stability property of the feasible set $M[F_1, F_2]$ w.r.t. C^1-perturbations of the defining functions F_1 and F_2. Under a C^1-neighborhood of a function $g \in C^1(\mathbb{R}^n, \mathbb{R}^l)$, we understand a subset of $C^1(\mathbb{R}^n, \mathbb{R}^l)$ that contains for some $\varepsilon > 0$ the set

$$\left\{ \widetilde{g} \in C^1(\mathbb{R}^n, \mathbb{R}^l) \,\middle|\, \sum_{i=1}^{l} \sup_{x \in \mathbb{R}^n} (|\widetilde{g}_i(x) - g_i(x)| + \|\nabla \widetilde{g}_i(x) - \nabla g_i(x)\|) \leq \varepsilon \right\}.$$

Definition 6. The feasible set $M[F_1, F_2]$ from (2.5) is called *locally stable* at $\bar{x} \in M[F_1, F_2]$ if there exists an $\mathbb{R}^n$-neighborhood V of $\bar{x}$ and a C^1-neighborhood U of (F_1, F_2) in $C^1(\mathbb{R}^n, \mathbb{R}^k) \times C^1(\mathbb{R}^n, \mathbb{R}^k)$ such that for every $(\widetilde{F}_1, \widetilde{F}_2) \in U$ the corresponding feasible set $M[\widetilde{F}_1, \widetilde{F}_2] \cap V$ is homeomorphic with $M[F_1, F_2] \cap V$.

Our main goal is to characterize the local stability property of the feasible set $M[F_1, F_2]$ in terms of the gradients of F_1 and F_2. In the case of standard nonlinear programming, (local) stability of the feasible set was studied in [32, 67] and is characterized by the MFCQ (see Theorem 4).

For stability in the MPCC setting we propose a kind of Mangasarian-Fromovitz condition (MFC) and its stronger version (SMFC). Section 2.2.1 will be devoted to the MFC and SMFC and their relations to other constraint qualifications (linear independence CQ, Mordukhovich's extremal principle, metric regularity, generalized Mangasarian-Fromovitz CQ, and standard subdifferential qualification condition). The conjectured equivalence of the MFC and SMFC is discussed. In Section 2.2.2, we prove that SMFC implies local stability and ensures that the MPCC feasible set is a Lipschitz manifold. Here, the application of nonsmooth versions of implicit function theorems (due to Clarke and Kummer) is crucial. We refer the reader to [19, 70] for details.

2.2.1 Constraint qualifications MFC and SMFC

Definitions of MFC and SMFC

Assume that Assumption A below holds throughout.

Assumption A *For every $\bar{x} \in M[F_1, F_2]$ and $i \in \{1, \ldots, k\}$, the set of vectors*

$$\{\nabla F_{j,i}(\bar{x}) \mid F_{j,i}(\bar{x}) = 0, j = 1, 2\}$$

is linearly independent.

Furthermore, we define for $\bar{x} \in M[F_1, F_2]$ and $i = 1, \ldots, k$ the (nonempty) convex hull

$$C_i(\bar{x}) := \text{conv}\{\nabla F_{j,i}(\bar{x}) \,|\, F_{j,i}(\bar{x}) = 0\}.$$

Note that $C_i(\bar{x}) = \partial \min\{F_{1,i}, F_{2,i}\}(\bar{x})$ is Clarke's subdifferential of the function $\min\{F_{1,i}(\cdot), F_{2,i}(\cdot)\}$ (see [13]).

Definition 7 (MFC and SMFC). The Mangasarian-Fromovitz condition (MFC) is said to hold at $\bar{x} \in M[F_1, F_2]$ if any k vectors $(w_1, \ldots, w_k) \in C_1(\bar{x}) \times \cdots \times C_k(\bar{x})$ are linearly independent.

The Strong Mangasarian-Fromovitz condition (SMFC) is said to hold at $\bar{x} \in M[F_1, F_2]$ if there exists a k-dimensional linear subspace E of $\mathbb{R}^n$ such that any k vectors $(u_1, \ldots, u_k) \in P_E(C_1(\bar{x})) \times \cdots \times P_E(C_k(\bar{x}))$ are linearly independent, where $P_E : \mathbb{R}^n \longrightarrow E$ denotes the orthogonal projection.

Remark 1. In the presence of additional C^1-equality and -inequality constraints in the description of the MPCC feasible set, the MFC will be enlarged by the standard MFCQ formulation with respect to these constraints.

We give some equivalent reformulations of the MFC and SMFC.

Lemma 1 (MFC and SMFC via Clarke's subdifferentials).

(a)   *The MFC at $\bar{x} \in \mathbb{R}^n$ means that Clarke's subdifferentials*

$$\partial \min\{F_{1,i}, F_{2,i}\}(\bar{x}), \ i = 1, \ldots, k \ \text{are linearly independent.}$$

(b)   *The SMFC holds at $\bar{x} \in \mathbb{R}^n$ if and only if there exists a basis decomposition of $\mathbb{R}^n$ given by a nonsingular $n \times n$ matrix A such that after the linear coordinate transformation $y := Ax$ Clarke's subdifferentials of the functions $h_i(y) := \min\{F_{1,i}(A^{-1}(y)), F_{2,i}(A^{-1}(y))\}$ w.r.t. $z := (y_{n-k+1}, \ldots, y_n)$ are linearly independent, i.e.*

$$\partial_z h_i(\bar{y}), \ i = 1, \ldots, k \ \text{are linearly independent,}$$

where $\partial_z h_i(\bar{y}) := \{\eta \in \mathbb{R}^k \,|\, \text{there exists } \xi \in \mathbb{R}^{n-k} \text{ with } [\xi, \eta] \in \partial h_i(\bar{y})\}$.

Proof. For (a), we only recall that $C_i(\bar{x}) = \partial \min\{F_{1,i}, F_{2,i}\}(\bar{x})$. To prove (b), we first calculate

$$\partial h_i(\bar{y}) = \partial \min\{F_{1,i}, F_{2,i}\}(\bar{x}) \cdot A^{-1} = C_i(\bar{x}) \cdot A^{-1}.$$

Hence, if the SMFC holds at $\bar{x}$, we take as columns of A^{-1} any orthogonal bases of $E^\perp$ and E. Conversely, given A, we set the linear subspace E to be spanned by the k last columns of A^{-1}. $\square$

Remark 2 (SMFC as a maximal rank condition). From Lemma 1 (b), we see that the SMFC is the so-called maximal rank condition (in terms of Clarke [13]) w.r.t. some basis decomposition of $\mathbb{R}^n$. It turns out that the concrete choice of such a basis decomposition may affect the validity of the maximal rank condition (see Example 11 for details). This means that the property of maximal rank is not basis-independent. This observation is crucial and motivates the SMFC (see also Section 2.2.2).

Lemma 2 (MFC and SMFC via basis enlargement).

(a)   The MFC holds at $\bar{x} \in \mathbb{R}^n$ if and only if for any $w_i \in C_i(\bar{x}), i = 1, \ldots, k$ there exist $\xi_1, \ldots, \xi_{n-k} \in \mathbb{R}^n$ such that the vectors $w_1, \ldots, w_k, \xi_1, \ldots, \xi_{n-k}$ are linearly independent.

(b)   The SMFC holds at $\bar{x} \in \mathbb{R}^n$ if and only if there exist $\xi_1, \ldots, \xi_{n-k} \in \mathbb{R}^n$ such that for any $w_i \in C_i(\bar{x}), i = 1, \ldots, k$ the vectors $w_1, \ldots, w_k, \xi_1, \ldots, \xi_{n-k}$ are linearly independent.

Proof. The proof of (a) follows immediately from the definition of linear independence. To prove (b), if the SMFC holds, we choose $\xi_1, \ldots, \xi_{n-k}$ as basis of $E^\perp$. Conversely, we set $E := (\operatorname{span}\{\xi_1, \ldots, \xi_{n-k}\})^\perp$ in the SMFC. $\square$

Furthermore, we notice that the MFC is a natural constraint qualification for the Clarke stationarity.

Definition 8 (Clarke stationarity; see [22, 105]). A point $\bar{x} \in M[F_1, F_2]$ is called Clarke stationary (C-stationary) if there exist real numbers $\lambda_{j,i}, j = 1, 2, i = 1, \ldots, k$ such that

$$\nabla f(\bar{x}) + \sum_{i=1}^{k} (\lambda_{1,i} \nabla F_{1,i}(\bar{x}) + \lambda_{2,i} \nabla F_{2,i}(\bar{x})) = 0,$$

$$F_{j,i}(\bar{x})\lambda_{j,i} = 0 \text{ for every } j = 1, 2, i = 1, \ldots, k,$$

$$\lambda_{1,i}\lambda_{2,i} \geq 0 \text{ for every } i \in \{1, \ldots, k\} \text{ with } F_{1,i}(\bar{x}) = F_{2,i}(\bar{x}) = 0.$$

Proposition 2 (MFC and C-stationarity). *If $\bar{x}$ is a local minimizer of the MPCC and MFC holds at $\bar{x}$, then $\bar{x}$ is C-stationary.*

Proof. Due to Lemma 1 in [105], if $\bar{x}$ is a local minimizer of the MPCC, then there exist real numbers $\lambda, \lambda_{j,i}, j = 1, 2, i = 1, \ldots, k$ (not all vanishing) such that

$$\lambda \nabla f(\bar{x}) + \sum_{i=1}^{k} (\lambda_{1,i} \nabla F_{1,i}(\bar{x}) + \lambda_{2,i} \nabla F_{2,i}(\bar{x})) = 0,$$

$$F_{j,i}(\bar{x})\lambda_{j,i} = 0 \text{ for every } j = 1, 2, i = 1, \ldots, k,$$

and

$$\lambda_{1,i}\lambda_{2,i} \geq 0 \text{ for every } i \in \{1, \ldots, k\} \text{ with } F_{1,i}(\bar{x}) = F_{2,i}(\bar{x}) = 0.$$

Clearly, if $\lambda = 0$, then the MFC is violated at $\bar{x}$. Hence, $\bar{x}$ is C-stationary. $\square$

For more details on C-stationarity and other stationarity concepts, such as W-, A-, M-, and S-stationarity, see [24], [88], [96], [105], [118], and Sections 2.2.1 and 2.4.

Conceptional relations to other CQ

We recall the well-known **LICQ for the MPCC** (e.g., [105, 106]), which is said to hold at $\bar{x} \in M[F_1, F_2]$ if

$\{\nabla F_{i,j}(\bar{x}) \,|\, F_{i,j}(\bar{x}) = 0, i = 1,\ldots,k, j = 1,2\}$ are linearly independent

The LICQ can be equivalently formulated in terms of the transversal intersection of stratified sets (see [63]). As shown in [106], the LICQ is a generic constraint qualification. However, the LICQ is not necessary for local stability, as one can see from Example 5. In this and all further examples, only the local stability in 0 is of interest.

Example 5 (2D, stable: one point $\longrightarrow$ one point).

The set $M^5 := \{(x,y) \in \mathbb{R}^2 \,|\, \min\{x,y\} = 0, \min\{x-y, 2x-y\} = 0\}$ is a singleton and is locally stable at 0 (see Figure 7). However, the LICQ does not hold at 0.

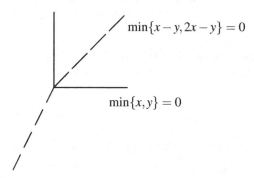

Figure 7 Illustration of Example 5

In this sense, the LICQ appears to be too restrictive. This comes from the fact that the LICQ does not impose the combinatorial structure of the complementarity constraints. Additionally, we notice that the LICQ implies the MFC.

Another condition we intend to discuss comes from the **exact Mordukhovich extremal principle** (see [25, 94]).

Let $\Omega \subset \mathbb{R}^n$ be any arbitrary closed set and $\bar{x} \in \Omega$. The nonempty cone

$$T(\bar{x}, \Omega) := \limsup_{\tau \searrow 0} \frac{\Omega - \bar{x}}{\tau}$$

$$= \left\{ d \in \mathbb{R}^n \,\Big|\, \text{there exist } x_k \longrightarrow \bar{x}, x_k \in \Omega, \tau_k \searrow 0 \text{ such that } \frac{x_k - \bar{x}}{\tau_k} \longrightarrow d \right\}$$

is called the contingent (also Bouligand or tangent) cone to Ω at x.

The Fréchet normal cone is defined via polarization as

$$\hat{N}(\bar{x}, \Omega) := (T(\bar{x}, \Omega))^\circ.$$

Finally, the limiting normal cone (also called the Mordukhovich normal cone) is defined by

$$N(\bar{x}, \Omega) := \limsup_{x' \xrightarrow{\Omega} \bar{x}} \hat{N}(x', \Omega)$$

$$= \left\{ \lim_{k \to \infty} w_k \,\middle|\, \text{there exist } x_k \longrightarrow \bar{x},\, x_k \in \Omega,\, w_k \in \hat{N}(x_k, \Omega) \right\}.$$

Definition 9 (local extremal point of set systems; [94]). Let Ω_i, $i = 1, \dots, k$ be nonempty subsets of $\mathbb{R}^n$ and $\bar{x} \in \bigcap_{i=1}^{k} \Omega_i$. We say that $\bar{x}$ is a local extremal point of the set system $\{\Omega_1, \dots, \Omega_k\}$ if there are sequences $\{a_{ij}\} \subset \mathbb{R}^n$, $i = 1, \dots, k$, and a neighborhood V of $\bar{x}$ such that $a_{ij} \longrightarrow 0$ as $j \longrightarrow \infty$ and

$$\bigcap_{i=1}^{k} (\Omega_i - a_{ij}) \cap V = \emptyset \text{ for all large } j \in \mathbb{N}.$$

We recall the finite-dimensional version of the exact Mordukhovich extremal principle.

Theorem 11 (Exact extremal principle in finite dimensions; [94]). *Let Ω_i, $i = 1, \dots, k$ be nonempty closed subsets of $\mathbb{R}^n$ and $\bar{x} \in \bigcap_{i=1}^{k} \Omega_i$ be an extremal point of the set system $\{\Omega_1, \dots, \Omega_k\}$. Then there are $x_i^* \in N(\bar{x}, \Omega_i)$, $i = 1, \dots, k$ (not all vanishing) such that $\sum_{i=1}^{k} x_i^* = 0$.*

Actually, Theorem 11 provides a sufficient condition for the property that the intersection of nonempty closed subsets Ω_i, $i = 1, \dots, k$ of $\mathbb{R}^n$ remains locally nonempty with respect to translations. This sufficient condition can be formulated as follows:

($\triangle$)   For all $x_i^* \in N(\bar{x}, \Omega_i)$, $i = 1, \dots, k$, $\sum_{i=1}^{k} x_i^* = 0$ implies $x_i^* = 0$, $i = 1, \dots, k$.

In order to refer to the foregoing discussion in our setting, from now on we set $\Omega_i := M_i$, $i = 1, \dots, k$, where

$$M_i := \{x \in \mathbb{R}^n \mid F_{1,i}(x) \geq 0, F_{2,i}(x) \geq 0, F_{1,i}(x)F_{2,i}(x) = 0\}.$$

Proposition 3 (MFC implies $\triangle$). *If the MFC holds at $\bar{x} \in M[F_1, F_2]$, then $\triangle$ also holds at $\bar{x}$.*

Proof. Let $i \in \{1, \dots, k\}$ be fixed. We provide a representation formula for $N(\bar{x}, M_i)$. We restrict ourselves to the interesting case that $F_{1,i}(\bar{x}) = F_{2,i}(\bar{x}) = 0$. Due to Assumption A, we choose vectors $\xi_1, \dots, \xi_{n-2} \in \mathbb{R}^n$, which form — together with the vectors $\nabla F_{1,i}(\bar{x}), \nabla F_{2,i}(\bar{x})$ — a basis for $\mathbb{R}^n$. Next we put $y = \Phi(x)$ as follows:

$$y_1 := F_{1,i}(x), \; y_2 := F_{2,i}(x), \; y_3 := \xi_1^T(x - \bar{x}), \dots, y_n := \xi_{n-2}^T(x - \bar{x}).$$

Note that $\Phi(\bar{x}) = 0$ and $D\Phi(\bar{x})$ is nonsingular. Therefore, Φ maps M_i diffeomorphically to $K := \{y \in \mathbb{R}^n \mid y_1 \geq 0, y_2 \geq 0, y_1 y_2 = 0\}$ locally at $\bar{x}$. Setting $L := \{y \in \mathbb{R}^2 \mid y_1 \geq 0, y_2 \geq 0, y_1 y_2 = 0\}$, Proposition 6.41 from [104] yields

$$N(0,K) = N(0, I_, \times \mathbb{R}^{n-2}) = N(0, I_,) \times N(0, \mathbb{R}^{n-2})$$

From [25] and [96], we conclude that $N(0,L) = \mathbb{R}^2_- \cup L$. Clearly, $N(0, \mathbb{R}^{n-2}) = \{0_{n-2}\}$. Altogether, we get

$$N(0,K) = \mathbb{R}^2_- \cup L \times \{0_{n-2}\}.$$

Using Exercise 6.7 (change of coordinates) from [104], we get

$$N(\bar{x}, M_i) = \{\beta_1 \nabla F_{1,i}(\bar{x}) + \beta_2 \nabla F_{2,i}(\bar{x}) \mid \text{either } \beta_1 < 0, \beta_2 < 0 \text{ or } \beta_1\beta_2 = 0\}. \quad (2.6)$$

Analogously, we obtain

$$\widehat{N}(\bar{x}, M_i) = \{\beta_1 \nabla F_{1,i}(\bar{x}) + \beta_2 \nabla F_{2,i}(\bar{x}) \mid \beta_1 \leq 0, \beta_2 \leq 0\}. \quad (2.7)$$

The representation (2.7) yields that the MFC is equivalent to the following condition:

For all $x_i^* \in \pm\widehat{N}(\bar{x}, M_i)$, $i = 1, \ldots, k$, $\sum_{i=1}^{k} x_i^* = 0$ implies $x_i^* = 0$, $i = 1, \ldots, k$.

Since $N(\bar{x}, M_i) \subset \pm\widehat{N}(\bar{x}, M_i)$, $i = 1, \ldots, k$, (see (2.6) and (2.7)), the proposition follows immediately. $\square$

Corollary 1 (MFC via Fréchet normal cones). *MFC is equivalent to the following condition:*

For all $x_i^* \in \pm\widehat{N}(\bar{x}, M_i)$, $i = 1, \ldots, k$, $\sum_{i=1}^{k} x_i^* = 0$ implies $x_i^* = 0$, $i = 1, \ldots, k$,

where $M_i = \{x \in \mathbb{R}^n \mid F_{1,i}(x) \geq 0, F_{2,i}(x) \geq 0, F_{1,i}(x)F_{2,i}(x) = 0\}$.

As we show by Example 6, $\triangle$ is not sufficient for $M[F_1, F_2]$ to be locally stable at 0. In this and all further examples in 3D, we understand under "two-star", "three-star" and "four-star" subsets of $\mathbb{R}^3$ as depicted in Figure 8 up to a homeomorphism.

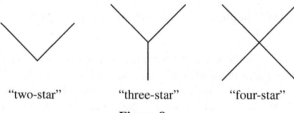

"two-star"　　　　　"three-star"　　　　　"four-star"

Figure 8

Example 6 (3D, nonstable: "four-star" $\longrightarrow$ 2 "two-stars").

Consider the "four-star" subset

$$M^6 := \{(x,y,z) \in \mathbb{R}^3 \mid \min\{x,y\} = 0, \min\{x+y-\sqrt{2}z, x+y+\sqrt{2}z\} = 0\}$$

(see Figure 9). After an appropriate perturbation the resulting set would have two path-connected components. Therefore, M^6 is not locally stable at 0.

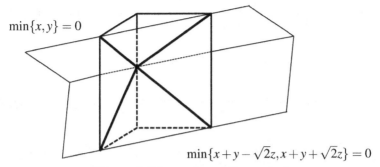

$\min\{x,y\} = 0$

$\min\{x+y-\sqrt{2}z, x+y+\sqrt{2}z\} = 0$

Figure 9 Illustration of Example 6

To show that $\triangle$ holds at 0, we set

$$M_1^6 := \{(x,y,z) \in \mathbb{R}^3 \mid \min\{x,y\} = 0\},$$

$$M_2^6 := \{(x,y,z) \in \mathbb{R}^3 \mid \min\{x+y-\sqrt{2}z, x+y+\sqrt{2}z\} = 0\},$$

and obtain due to (2.6) from the proof of Proposition 3

$$N(0,M_1^6) = \{(\beta_1,\beta_2,0)^T \in \mathbb{R}^3 \mid \text{either } \beta_1 < 0, \beta_2 < 0 \text{ or } \beta_1\beta_2 = 0\}.$$

$$N(0,M_2^6) = \left\{ \beta_1 \begin{pmatrix} 1 \\ 1 \\ -\sqrt{2} \end{pmatrix} + \beta_2 \begin{pmatrix} 1 \\ 1 \\ \sqrt{2} \end{pmatrix} \middle| \text{ either } \beta_1 < 0, \beta_2 < 0 \text{ or } \beta_1\beta_2 = 0 \right\}.$$

From the representations of $N(0,M_1^6)$ and $N(0,M_2^6)$, it is easy to see that $\triangle$ (but not MFC) is satisfied at $0 \in M^6$.

The next stability concept we would like to discuss here is **metric regularity**. We recall that a multivalued map $T : \mathbb{R}^n \rightrightarrows \mathbb{R}^k$ is called metrically regular (with rank $L > 0$) at $(\bar{x},\bar{y}) \in \text{gph } T$ if, for certain neighborhoods U and V of $\bar{x}$ and $\bar{y}$, respectively, it holds that

$$\text{dist}(x, T^{-1}(y)) \leq L\,\text{dist}(y, T(x)) \text{ for all } x \in U, y \in V.$$

Furthermore, a multivalued map $S : \mathbb{R}^k \rightrightarrows \mathbb{R}^n$ is called pseudo-Lipschitz (with rank $L > 0$) at $(\bar{y},\bar{x}) \in \text{gph}S$ if there are neighborhoods U and V of $\bar{x}$ and $\bar{y}$, respectively, such that, given any points $(y,x) \in (V \times U) \cap \text{gph}S$, it holds that

$$\text{dist}(x, S(y')) \leq L\|y' - y\| \text{ for all } y' \in V$$

(see, e.g., [49], [81]).

It holds (see [48]) that T is metrically regular at $(\bar{x},\bar{y}) \in \text{gph } T$ if and only if T^{-1} is pseudo-Lipschitz at $(\bar{y},\bar{x})$.

It is well-known from [100] that the solution map

$$S(y,z) := \{x \in \mathbb{R}^n \mid h(x) = y, g(x) \leq z\}, \ (g,h) \in C^1(\mathbb{R}^n, \mathbb{R}^{k+m}),$$

is pseudo-Lipschitz at $(0,0,\bar{x})$ if and only if the MFCQ is satisfied at $\bar{x} \in S(0,0)$. This means that the local stability of $M_{NLP}[h,g] \ (= S(0,0))$ at $\bar{x} \in M_{NLP}[h,g]$ is equivalent to the metric regularity of

$$S^{-1}(x) = \{(h(x),z) \mid g(x) \leq z\}$$

at $(\bar{x},0,0)$.

To apply this idea in our setting, we say that the metric regularity condition (MRC) holds at $\bar{x} \in M[F_1,F_2]$ if and only if

$$G : \begin{cases} \mathbb{R}^n \longrightarrow \mathbb{R}^k, \\ x \ \mapsto \ (\min\{F_{1,i}(x), F_{2,i}(x)\})_{i=1,\ldots,k}, \end{cases}$$

is metrically regular at $(\bar{x},0)$.

The next proposition can be derived with the aid of Proposition 3.3 in [52]. For the sake of completeness, we present its proof.

Proposition 4 (MRC is equivalent to $\triangle$). *MRC holds at $\bar{x} \in M[F_1,F_2]$ if and only if $\triangle$ holds at $\bar{x}$.*

Proof. The MRC holds at $\bar{x} \in M[F_1,F_2]$ if and only if the solution map $S(y) := \{x \in \mathbb{R}^n \mid G(x) = y\}$, $y \in \mathbb{R}^k$, is pseudo-Lipschitz at $(0,\bar{x})$. Setting

$$F : \begin{cases} \mathbb{R}^n \longrightarrow \mathbb{R}^{2k}, \\ x \ \mapsto \ (F_{1,i}(x), F_{2,i}(x))_{i=1,\ldots,k}, \end{cases}$$

and $D_i := \{(a_i,b_i) \in R^2 \mid a_i \geq 0, \ b_i \geq 0, \ ab = 0\}$, $i = 1,\ldots,k$, we obtain:

$$S(y) = \{x \in \mathbb{R}^n \mid F(x) - y \in D_1 \times \cdots \times D_k\},$$

$$S^{-1}(x) = F(x) - D_1 \times \cdots \times D_k.$$

Therefore, the MRC holds at $\bar{x} \in M[F_1,F_2]$ if and only if $F(\cdot) - D_1 \times \cdots \times D_k$ is metrically regular at $(\bar{x},0)$. Since $F \in C^1(\mathbb{R}^n, \mathbb{R}^{2k})$ and $D_1 \times \cdots \times D_k$ is closed, we can apply Example 9.44 from [104]. Due to that example the constraint qualification

$$u \in N(F(\bar{x}), D_1 \times \cdots \times D_k), \ \nabla^T F(\bar{x})u = 0 \Longrightarrow u = 0, \qquad (2.8)$$

is equivalent to the metric regularity of $F(\cdot) - D_1 \times \cdots \times D_k$ at $(\bar{x},0)$. Since

$$N(F(\bar{x}), D_1 \times \cdots \times D_k) = N(F_{1,1}(\bar{x}), F_{2,1}(\bar{x}), D_1) \times \cdots \times N(F_{1,k}(\bar{x}), F_{2,k}(\bar{x}), D_k)$$

and

$$N(0,D_i) = \mathbb{R}^2_- \cup D_i,$$

formula (2.6) allows to conclude that the constraint qualification (2.8) is equivalent to $\triangle$. $\square$

We mention some valuable remarks on the previously discussed constraint qualifications (pointed out by an anonymous referee).

Remark 3 (Standard subdifferential qualification condition $\triangle$). $\triangle$ is the standard subdifferential qualification condition for the system Ω_i, $i = 1, \ldots, k$ at $\bar{x} \in \bigcap_{i=1}^{k} \Omega_i$ (see [52, 94, 104]). Moreover, $\triangle$ means that the multivalued map $M(z) := \{x \in \mathbb{R}^n \,|\, x + z_i \in \Omega_i, i = 1, \ldots, k\}, z = (z_1, \ldots, z_k) \in \mathbb{R}^{nk}$, is pseudo-Lipschitz (has the Aubin property) at $(0, \ldots, 0, \bar{x})$. This means that its inverse, $M^{-1}(x) := (\Omega_1 - x) \times \ldots \times (\Omega_1 - x), x \in \mathbb{R}^n$, is metrically regular at $(\bar{x}, 0, \ldots, 0)$ (e.g., Proposition 3.3 in [52]).

Remark 4 (MFC and GMFCQ). The generalized Mangasarian-Fromovitz constraint qualification (GMFCQ) can be related to the MFC. Indeed, the GMFCQ for the constraint set $M = \{x \in \mathbb{R}^n \,|\, F(x) \in D_1 \times \cdots \times D_k\}$ is exactly (2.8) (see [52]). Thus, it is clear from the proof of Proposition 4 that $\triangle$ is equivalent to the GMFCQ. Hence, the MFC implies the MRC, as well as the GMFCQ. Moreover, Example 6 shows that neither the MRC nor the GMFCQ is sufficient for $M[F_1, F_2]$ being locally stable.

Remark 5 (Constraint qualifications for M-stationarity). It is well-known that under $\triangle$ (or, equivalently, the MRC and GMFCQ) a local minimum for (2.5) is M-stationary. This means that in addition to C-stationarity in Definition 8 it holds that either $\lambda_{1,i}, \lambda_{2,i} < 0$ or $\lambda_{1,i}\lambda_{2,i} = 0$ for every $i \in \{1, \ldots, k\}$ with $F_{1,i}(\bar{x}) = F_{2,i}(\bar{x}) = 0$.

On equivalence of MFC and SMFC

It is clear that the SMFC implies the MFC. Moreover, these two conditions coincide for $n = k$. The question of whether the SMFC is equivalent to the MFC in general is highly nontrivial.

First, we show that the SMFC implies the MFC at least in the cases where $k = 2$ or LICQ is fulfilled.

This follows mainly from the (linear-algebraic) Lemma 3.

Lemma 3. *Let* $C_i := conv\{v_{j,i} \in \mathbb{R}^n \,|\, j = 1, 2\}$, $i = 1, \ldots, k$, *and for every* $i \in \{1, \ldots, k\}$ *let* $v_{1,i}, v_{2,i}$ *be linearly independent. Let assertions (A) and (B) be given as follows:*

(A) *Any* k *vectors* $(w_1, \ldots, w_k) \in C_1 \times \cdots \times C_k$ *are linearly independent.*
(B) *There exists a* k-*dimensional linear subspace* E *of* $\mathbb{R}^n$ *such that any* k *vectors* $(u_1, \ldots, u_k) \in P_E(C_1) \times \cdots \times P_E(C_k)$ *are linearly independent, where* $P_E : \mathbb{R}^n \longrightarrow E$ *denotes the orthogonal projection.*

Then, (A) and (B) are equivalent in the following cases:

1) *The vectors* $v_{j,i}$, $j = 1, 2$, $i = 1, \ldots, k$ *are linearly independent.*
2) $k = 2$.

Proof. The nontrivial part is to prove that (A) implies (B) for $n > k$. First, we claim that (B) is equivalent to the following condition (C) (see Lemma 2):

(C) There exist $\xi_1, \ldots, \xi_{n-k} \in \mathbb{R}^n$ such that for any $w_i \in C_i, i = 1, \ldots, k$ the vectors $w_1, \ldots, w_k, \xi_1, \ldots, \xi_{n-k}$ are linearly independent.

Indeed, if (B) holds, we choose $\xi_1, \ldots, \xi_{n-k}$ as a basis of $E^{\perp}$ in (C). If (C) holds, we set $E := (\text{span}\{\xi_1, \ldots, \xi_{n-k}\})^{\perp}$ in (B).

Case 1: the vectors $v_{j,i}, j = 1, 2, i = 1, \ldots, k$ are linearly independent.

Then, $n \geq 2k$ and $v_{j,i}, j = 1, 2, i = 1, \ldots, k$ span a $2k$-dimensional linear subspace of $\mathbb{R}^n$. Hence, w.l.o.g., we may assume that $n = 2k$.

Define a linear coordinate transformation $L : \mathbb{R}^n \longrightarrow \mathbb{R}^n$ as

$$L(v_{1,i}) = e_{2i-1} + e_{2i}, L(v_{2,i}) = e_{2i-1}, i = 1, \ldots, k,$$

whereby e_m denotes the m-th standard basis vector for $1 \leq m \leq n$. It holds that $L(C_i) = \{e_{2i-1} + \lambda_i e_{2i} \mid \lambda_i \in [0,1]\}, i = 1, \ldots, k$.

Setting $T := \text{span}\{e_{2i-1}, i = 1, \ldots, k\}$, we obviously obtain that

$(\star)$ any k vectors $(v_1, \ldots, v_k) \in P_T(L(C_1)) \times \cdots \times P_T(L(C_k))$ are linearly independent, where $P_T : \mathbb{R}^n \longrightarrow T$ denotes the orthogonal projection.

As above, $(\star)$ is equivalent to the following condition:

$(\star\star)$ There exist $\gamma_1, \ldots, \gamma_{n-k} \in \mathbb{R}^n$ such that for any $v_i \in L(C_i), i = 1, \ldots, k$, the vectors $v_1, \ldots, v_k, \gamma_1, \ldots, \gamma_{n-k}$ are linearly independent.

Setting $\xi_i := L^{-1}(\gamma_i), i = 1, \ldots, n-k$, we conclude that (C) is fulfilled due to $(\star\star)$. Thus, B is proved.

Case 2: $k = 2$.

It is clear that the vectors $v_{j,i}, j = 1, 2, i = 1, 2$ span at most a *four*-dimensional linear subspace S of $\mathbb{R}^n$ and hence $\dim S \leq 4$. If $\dim S = 4$, then (B) holds as in Case 2. If $\dim S < 4$, we may assume w.l.o.g. that $n = 3$.

For $a \in \{-1, 1\}^2$, we set $K_a := \text{cone}\{a_i v_{1,i}, a_i v_{2,i} \mid i = 1, 2\}$. From the theorem about alternatives (e.g., [102]), we claim that (A) is equivalent to the following condition:

$$\text{int}(K_a^{\circ}) \neq \emptyset \text{ for all } a \in \{-1, 1\}^2.$$

Here, $\text{int}(K_a^{\circ})$ denotes the interior of the polar cone of K_a.

Due to this fact, K_a properly lies in a half-space for all $a \in \{-1, 1\}^2$. Setting $\{-1, 1\}^2 =: \{a^1, -a^1, a^2, -a^2\}$, we can strictly separate K_{a^l} and K_{-a^l} by a plane $\beta_l \ni 0, l = 1, 2$. Since $0 \in \beta_1 \cap \beta_2$, there exists $\xi \in \beta_1 \cap \beta_2, \xi \neq 0$, such that $\xi \notin \bigcup_{a \in \{-1,1\}^2} K_a$ by construction. This means that (C) is fulfilled. Thus, (B) is proved. $\square$

Theorem 12 (MFC implies SMFC for $k = 2$ and under LICQ). *Let $k = 2$ or the LICQ be fulfilled. Then, the SMFC is equivalent to the MFC.*

Proof. It is straightforward to see that the conclusion can be obtained by applying Lemma 3. We have to adjust the proof of Lemma 3 only for the case where only one constraint in $\min\{F_{1,i}(\bar{x}), F_{2,i}(\bar{x})\} = 0$ is active (i.e., $F_{1,i}(\bar{x}) = 0, F_{2,i}(\bar{x}) > 0$, or vice versa). For that we define C_i from Lemma 3 just to be $C_i(\bar{x})$. The respective change in the proof of Lemma 3 is straightforward. In fact, only the so-called biactive set of constraints is crucial (see [88], [118]). $\square$

Remark 6 (MFC implies SMFC for $k = 3$; Rückmann, personal communication). Recently it was proven that the MFC implies the SMFC when $k = 3$. The proof uses a kind of dual description of the SMFC and MFC.

In what follows, we discuss the difficulties by proving that the MFC implies the SMFC for general n and k. These difficulties arise not so much because of linear-algebraic issues but rather because of combinatorial and topological matters of the problem. In fact, using the notation from Lemma 3, we set for $a \in \{-1, 1\}^k$

$$K_a := \text{cone}\{a_i v_{1,i}, a_i v_{2,i} \mid i = 1, \ldots, k\}.$$

Condition (A) means that all cones K_a are pointed; that is,

$$\text{if } x_1 + \cdots + x_p = 0, x_s \in K_a, s = 1, \ldots, p, \text{ then } x_s = 0 \text{ for all } s = 1, \ldots, p.$$

Condition (B) means that there exist $n - k$ linearly independent vectors $\xi_1, \ldots, \xi_{n-k} \in \mathbb{R}^n$ such that

$$\xi_j \notin \bigcup_{a \in \{-1,1\}^k} K_a \text{ for all } j = 1, \ldots, n - k.$$

Thus, for proving "(A) implies (B)", we need to show that (for all k and n)

$$\bigcup_{a \in \{-1,1\}^k} K_a \neq \mathbb{R}^n. \tag{2.9}$$

Here, we deal with a union of pointed cones with the additional property that

$$K_{-a} = -K_a \text{ for all } a \in \{-1, 1\}^k.$$

Moreover, 2^k — the number of these cones — grows exponentially in k. It is clear that for proving (2.9) topological properties of $\bigcup_{a \in \{-1,1\}^k} K_a$ (such as, for example, Euler characteristic) are crucial.

We conclude by noting that the conjectured equivalence of the MFC and SMFC is very sophisticated and is a topic of current research.

2.2.2 SMFC implies stability and Lipschitz manifold

We intend to prove that the SMFC implies local stability of the feasible set $M[F_1,F_2]$ (see Theorem 15). The main idea is to show that under the SMFC $M[F_1,F_2]$ appears to be an $(n-k)$-dimensional Lipschitz manifold (the Corollary 2 and Definition 1).

Guiding examples

First, we briefly mention two- and three-dimensional examples with two linear constraints. These examples illustrate which phenomena might occur in general. They mainly highlight the possibilities arising with respect to the stability property of the feasible set $M[F_1,F_2]$ in low dimensions.

Example 7 (2D, nonstable: one point $\longrightarrow$ empty, two points). The set $M^7 := \{(x,y) \in \mathbb{R}^2 \mid \min\{x,y\} = 0, \min\{-x,-y\} = 0\}$ is a singleton (see Figure 10). Note that MFC is not satisfied at 0. After an appropriate perturbation M^7 either becomes empty or contains at least two points.

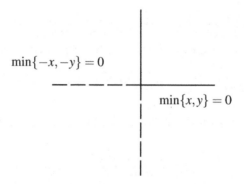

Figure 10 Illustration of Example 7

Example 8 (2D, nonstable: one point $\longrightarrow$ two points). The set $M^8 := \{(x,y) \in \mathbb{R}^2 \mid \min\{x,y\} = 0, \min\{-x+y,x+y\} = 0\}$ is a singleton (see Figure 11). Note that the MFC is not satisfied at 0. After an appropriate perturbation, M^8 contains at least two points.

$$\min\{-x+y, x+y\} = 0$$

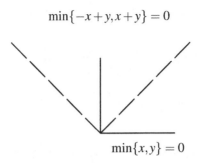

$$\min\{x, y\} = 0$$

Figure 11 Illustration of Example 8

Example 9 (3D, nonstable: "three-star" $\longrightarrow$ *1 or 2 "two-stars").* The set $M^9 :=$ $\{(x,y,z) \in \mathbb{R}^3 \,|\, \min\{x,y\} = 0, \min\{y-z,y+z\} = 0\}$ is a "three-star" (see Figure 12). Note that the MFC is not satisfied at 0. After an appropriate perturbation, M^9 either has two path-connected components or is a "two-star".

$$\min\{x, y\} = 0$$

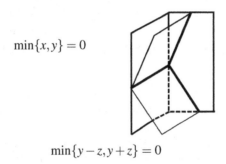

$$\min\{y-z, y+z\} = 0$$

Figure 12 Illustration of Example 9

Example 10 (3D, stable: "two-star" $\longrightarrow$ *"two-star").* The set $M^{10} := \{(x,y,z) \in \mathbb{R}^3 \,|\, \min\{x,y\} = 0, \min\{x-y+z, -x+y+z\} = 0\}$ is a "two-star"(see Figure 13). Note that the MFC holds at 0. After any sufficiently small perturbation, M^{10} remains to be a "two-star".

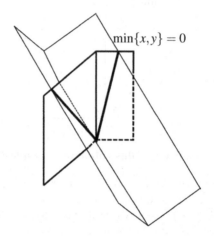

$$\min\{x-y+z,-x+y+z\} = 0$$

Figure 13 Illustration of Example 10

It is easy to see that in all these examples the MFC holds at 0 if and only if the corresponding feasible set is locally stable. Moreover, these examples emphasize that the locally stable case corresponds to a feasible set being a Lipschitz manifold (see Corollary 2 below).

Main results via Clarke's implicit function theorem

We recall briefly the notion of Clarke's generalized Jacobian and the corresponding inverse and implicit function theorems (see [13] and Section B.1).

For a vector-valued function $G = (g_1,\ldots,g_k) : \mathbb{R}^n \longrightarrow \mathbb{R}^k$ with g_i being Lipschitz near $\bar{x} \in \mathbb{R}^n$, the set

$$\partial G(\bar{x}) := \text{conv}\{\lim DG(x_i) \,|\, x_i \longrightarrow \bar{x}, x_i \notin \Omega_G\}$$

is called Clarke's generalized Jacobian, where $\Omega_G \subset \mathbb{R}^n$ denotes the set of points at which G fails to be differentiable.

Theorem 13 (Clarke's inverse function theorem [13]). *Let $F : \mathbb{R}^n \longrightarrow \mathbb{R}^n$ be Lipschitz near $\bar{x}$. If all matrices in $\partial F(\bar{x})$ are nonsingular, then F has the unique Lipschitz inverse function F^{-1} locally around $\bar{x}$.*

Theorem 14 (Clarke's implicit function theorem [13]). *Let $G : \mathbb{R}^{n-k} \times \mathbb{R}^k \longrightarrow \mathbb{R}^k$ be Lipschitz near $(\bar{y},\bar{z}) \in \mathbb{R}^{n-k} \times \mathbb{R}^k$ with $G(\bar{y},\bar{z}) = 0$. Suppose that*

$$\pi_z \partial G(\bar{y},\bar{z}) := \{M \in \mathbb{R}^{k \times k} \,|\, \text{there exists } N \in \mathbb{R}^{k \times n} \text{ with } [N,M] \in \partial G(\bar{y},\bar{z})\}$$

is of maximal rank (i.e., contains merely nonsingular matrices). Then there exist an $\mathbb{R}^{n-k}$-neighborhood Y of $\bar{y}$, an $\mathbb{R}^k$-neighborhood Z of $\bar{z}$, and a Lipschitz function

$\zeta : Y \longrightarrow Z$ such that $\zeta(\bar{y}) = \bar{z}$ and for every $(y,z) \in Y \times Z$ it holds that

$$G(y,z) = 0 \text{ if and only if } z = \zeta(y).$$

However, Example 11 illustrates that Theorem 14 cannot be applied directly in general just for the linear case of a stable $M[F_1, F_2]$.

Example 11 (3D, stable: IFT is not applicable). Consider the set $M^{11} := \{(x,y,z) \in \mathbb{R}^3 \mid \min\{x,y\} = 0, \min\{-y+z,z\} = 0\}$ (see Figure 14). This example shows that although $M[F_1, F_2]$ is a Lipschitz manifold, it cannot be parameterized by means of any splitting of $\mathbb{R}^3$ in the *standard* basis. Therefore, Theorem 14 (and, actually, any implicit function theorem) cannot be applied directly.

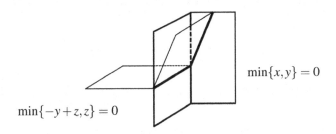

$\min\{x,y\} = 0$

$\min\{-y+z,z\} = 0$

Figure 14 Illustration of Example 11

Indeed, Example 11 suggests first performing a linear coordinate transformation in order to make Theorem 14 applicable. Exactly this idea is incorporated in the SMFC and allows us to prove the following result.

Theorem 15 (Local stability under SMFC). *If the SMFC holds at $x \in M[F_1, F_2]$, then the feasible set $M[F_1, F_2]$ is locally stable at $\bar{x}$.*

Proof. Let $\bar{x} \in M[F_1, F_2]$. Since the SMFC holds at $\bar{x}$, there exists a k-dimensional linear subspace E of $\mathbb{R}^n$ such that any k vectors $(u_1, \ldots, u_k) \in P_E(C_1(\bar{x})) \times \cdots \times P_E(C_k(\bar{x}))$ are linearly independent. Without loss of generality, we may assume that $E = \{0_{n-k}\} \times \mathbb{R}^k$.

Setting $g_i := \min\{F_{1,i}, F_{2,i}\}$, $i = 1, \ldots, k$, we define

$$G : \begin{cases} \mathbb{R}^{n-k} \times \mathbb{R}^k \longrightarrow \mathbb{R}^k, \\ (y,z) \mapsto (g_1(y,z), \ldots, g_k(y,z)). \end{cases}$$

Let $\bar{x} = (\bar{y}, \bar{z}) \in \mathbb{R}^{n-k} \times \mathbb{R}^k$. We obtain from $\partial g_i(\bar{x}) = C_i(\bar{x})$, $i = 1, \ldots, k$, and the choice of E that

$$\pi_z \partial G(\bar{y}, \bar{z}) \subset P_E(C_1(\bar{x})) \times \cdots \times P_E(C_k(\bar{x})).$$

Hence, from the SMFC, $\pi_z \partial G(\bar{y}, \bar{z})$ is of maximal rank and Theorem 14 can be applied. Then there exist a compact $\mathbb{R}^{n-k}$-neighborhood Y of $\bar{y}$, an $\mathbb{R}^k$-neighborhood

Z of $\bar{z}$ and a Lipschitz function $\zeta : Y \longrightarrow Z$ such that $\zeta(\bar{y}) = \bar{z}$ and for every $(y, z) \in Y \times Z$ it holds that

$$G(y,z) = 0 \text{ if and only if } z = \zeta(y).$$

For $\varepsilon > 0$, we set

$$K_\varepsilon := (\{(y, \zeta(y)) \mid y \in Y\} + \bar{B}_{\mathbb{R}^n}(0, \varepsilon)) \cap (Y \times Z),$$

an ε-tube around $M[F_1, F_2] \cap (Y \times Z)$. Due to the compactness of Y, continuity reasonings, and stability of the SMFC within the space of C^1-functions (taking Y smaller if needed), there exists $\varepsilon > 0$ such that:

($\bullet$)　　$K_\varepsilon \subset Y \times Z$ and K_ε is compact.
($\bullet\bullet$)　　There exists a C^1-neighborhood U of (F_1, F_2) in $C^1(\mathbb{R}^n, \mathbb{R}^k) \times C^1(\mathbb{R}^n, \mathbb{R}^k)$ such that for every $(\widetilde{F}_1, \widetilde{F}_2) \in U$ it holds that

$$M[\widetilde{F}_1, \widetilde{F}_2] \cap (Y \times Z) \subset K_\varepsilon.$$

We assume U to be a ball of radius $r > 0$ in $C^1(\mathbb{R}^n, \mathbb{R}^k) \times C^1(\mathbb{R}^n, \mathbb{R}^k)$.
($\bullet\bullet\bullet$)　　The SMFC is fulfilled at every $x \in M[\widetilde{F}_1, \widetilde{F}_2] \cap (Y \times Z)$ for every $(\widetilde{F}_1, \widetilde{F}_2) \in U$ with the same k-dimensional linear subspace E.

Now $(\widetilde{F}_1, \widetilde{F}_2) \in U$ be arbitrary but fixed. Setting $\widetilde{g}_i := \min\{\widetilde{F}_{1,i}, \widetilde{F}_{2,i}\}$, $i = 1, \ldots, k$, we define

$$\widetilde{G} : \begin{cases} \mathbb{R}^{n-k} \times \mathbb{R}^k \longrightarrow & \mathbb{R}^k, \\ (y, z) \mapsto & (\widetilde{g}_1(y, z), \ldots, \widetilde{g}_k(y, z)). \end{cases}$$

Our aim is to show that for every fixed $\widetilde{y} \in Y$ the equation $\widetilde{G}(\widetilde{y}, z) = 0$ is uniquely solvable with $(\widetilde{y}, z) \in K_\varepsilon$. For that, we set for $(t, y, z) \in [0, 1] \times \mathbb{R}^{n-k} \times \mathbb{R}^k$

$$H_{1,i}(t, y, z) := (1-t)F_{1,i}(y, z) + t\widetilde{F}_{1,i}(y, z),$$

$$H_{2,i}(t, y, z) := (1-t)F_{2,i}(y, z) + t\widetilde{F}_{2,i}(y, z),$$

$$g_i(t, y, z) := \min\{H_{1,i}(t, z), H_{2,i}(t, z)\}.$$

Furthermore, we construct a homotopy mapping

$$H : \begin{cases} [0, 1] \times \mathbb{R}^{n-k} \times \mathbb{R}^k \longrightarrow & \mathbb{R}^k, \\ (t, y, z) \mapsto & (g_1(t, y, z), \ldots, g_k(t, y, z)). \end{cases}$$

We keep in mind that $H(0, y, z) = G(y, z)$ and $H(1, y, z) = \widetilde{G}(y, z)$; moreover,

$$(H_1(t, \cdot, \cdot), H_2(t, \cdot, \cdot)) \in U \text{ for every } t \in [0, 1].$$

Next, we fix $\widetilde{y} \in Y$ and consider the equation $H(t, \widetilde{y}, z) = 0$ near its solution $(0, \widetilde{y}, \zeta(\widetilde{y}))$. Since $(\widetilde{y}, \zeta(\widetilde{y})) \in M[F_1, F_2] \cap (Y \times Z)$, we obtain from ($\bullet\bullet\bullet$) that the SMFC holds at $(\widetilde{y}, \zeta(\widetilde{y}))$. This means that

$$\pi_z \partial H(0, \widetilde{y}, \zeta(\widetilde{y})) = \pi_z \partial G(\widetilde{y}, \zeta(\widetilde{y}))$$

is of maximal rank and Theorem 14 can be applied for $H(t, \widetilde{y}, z) = 0$ near its solution $(0, \widetilde{y}, \zeta(\widetilde{y}))$. Thus, we obtain for every $t \in [0, \delta), 0 < \delta \leq 1$ a solution $z(t)$ such that $H(t, \widetilde{y}, z(t)) = 0$. Since $(H_1(t, \cdot, \cdot), H_2(t, \cdot, \cdot)) \in U$, $(\bullet\bullet)$ yields that $(\widetilde{y}, z(t)) \in K_\varepsilon$ for every $t \in [0, \delta]$. Here, δ is taken smaller if needed.

These considerations allow us to claim that

$$\widetilde{t} := \sup\{\bar{t} \in [0, 1) \mid \text{for every } t \in [0, \bar{t}) \text{ there exists at least one}$$
$$(\widetilde{y}, z(t)) \in K_\varepsilon \text{ such that } H(t, \widetilde{y}, z(t)) = 0\}$$

is well-defined.

Assume that $\widetilde{t} \neq 1$. Then, there is a sequence of solutions $z(t_m), t_m \in [0, \widetilde{t}), t_m \longrightarrow \widetilde{t}$ such that $(\widetilde{y}, z(t_m)) \in K_\varepsilon$ and $H(t_m, \widetilde{y}, z(t_m)) = 0$. We use the compactness of K_ε from $(\bullet)$ to obtain the existence of $\widetilde{z}$ with $(\widetilde{y}, \widetilde{z}) \in K_\varepsilon$ and $z_m \longrightarrow \widetilde{z}$. Hence, due to the continuity, we get in the limit $H(\widetilde{t}, \widetilde{y}, \widetilde{z}) = 0$. This conclusion allows us to apply Theorem 14 for the equation $H(t, \widetilde{y}, z) = 0$ near $(\widetilde{t}, \widetilde{y}, \widetilde{z})$ to extend the solution for $t > \widetilde{t}$. This yields a contradiction with the definition of $\widetilde{t}$.

So, we claim that $\widetilde{t} = 1$ and as above we obtain that $\widetilde{G}(\widetilde{y}, z) \equiv H(1, \widetilde{y}, z) = 0$ is solvable with $(\widetilde{y}, z) \in K_\varepsilon$.

The unique solvability of $\widetilde{G}(\widetilde{y}, z) = 0$ for $(\widetilde{y}, z) \in K_\varepsilon$ can be proven by contradiction using analogous arguments. One has only to follow different solutions by applying Theorem 14 successively until the unique solution $(0, \widetilde{y}, \zeta(\widetilde{y}))$ of $G(\widetilde{y}, z) \equiv H(t, \widetilde{y}, z) = 0$ is reached.

From all of this, it is proven that for every $\widetilde{y} \in Y$ the equation $\widetilde{G}(\widetilde{y}, z) = 0$ is uniquely solvable with $(\widetilde{y}, z(\widetilde{y})) \in K_\varepsilon$. From $(\bullet\bullet)$, one can immediately see that $\widetilde{G}(\widetilde{y}, z) = 0$ is uniquely solvable, actually, in Z. Therefore, $M[\widetilde{F}_1, \widetilde{F}_2] \cap (Y \times Z) = \{(y, z(y)) \mid y \in Y\}$. Here, $z : Y \longrightarrow Z$ is Lipschitz due to $(\bullet\bullet\bullet)$ and Theorem 14, which is applicable locally around every $\widetilde{x} \in M[\widetilde{F}_1, \widetilde{F}_2] \cap (Y \times Z)$.

It remains to add that $M[F_1, F_2] \cap (Y \times Z)$ and $M[\widetilde{F}_1, \widetilde{F}_2] \cap (Y \times Z)$, both being Lipschitz graphs on Y, are homeomorphic with $\mathbb{R}^{n-k}$ and thus with each other.□

From the proof of Theorem 15, we deduce the following Corollary 2.

Corollary 2. *If the SMFC holds at every $\bar{x} \in M[F_1, F_2]$, then the feasible set $M[F_1, F_2]$ is an $(n-k)$-dimensional Lipschitz manifold.*

Proof. We use the notation in Theorem 15. From the SMFC at $\bar{x} \in M[F_1, F_2]$, we may assume that after an appropriate linear coordinate transformation it holds that

$$(Y \times Z) \cap M[F_1, F_2] = \{(y, \zeta(y)) \mid y \in Y\},$$

where $\bar{x} = (\bar{y}, \bar{z}) \in \mathbb{R}^{n-k} \times \mathbb{R}^k$, Y is an $\mathbb{R}^{n-k}$-neighborhood of $\bar{y}$, Z is an $\mathbb{R}^k$-neighborhood of $\bar{z}$, and $\zeta : Y \longrightarrow Z$ is Lipschitz. Hence, $M[F_1, F_2]$ is locally the graph of a Lipschitz function ζ. $M[F_1, F_2]$ fits Definition 1 and is an $(n-k)$-dimensional Lipschitz manifold.□

On application of Kummer's implicit function theorem

In this section, we link the SMFC with the so-called Thibault limiting sets (or strict graphical derivatives) via Kummer's implicit function theorem (cf. [85] and Section B.1).

For a vector-valued function $G = (g_1, \ldots, g_k) : \mathbb{R}^n \longrightarrow \mathbb{R}^k$, the mapping $TG(\bar{x}) : \mathbb{R}^n \longrightarrow \mathbb{R}^k$ with

$$TG(\bar{x})(\bar{u}) := \left\{ v \in \mathbb{R}^k \; \middle| \; \begin{array}{l} v = \lim\limits_{k \to \infty} \dfrac{f(x_k + t_k u_k) - f(x_k)}{t_k} \\ \text{for certain } t_k \downarrow 0, \, x_k \longrightarrow \bar{x}, u_k \longrightarrow \bar{u} \end{array} \right\}$$

is called the Thibault derivative at $\bar{x}$ (see [114, 115]) or strict graphical derivative (see [104]).

If, additionally, g_i are Lipschitz near $\bar{x} \in \mathbb{R}^n$, then we may omit the sequence $u_k \longrightarrow \bar{u}$ in the definition of $TG(\bar{x})(\bar{u})$ and we get

$$TG(\bar{x})(\bar{u}) = \left\{ v \in \mathbb{R}^k \; \middle| \; \begin{array}{l} v = \lim\limits_{k \to \infty} \dfrac{f(x_k + t_k \bar{u}) - f(x_k)}{t_k} \\ \text{for certain } t_k \downarrow 0, \, x_k \longrightarrow \bar{x} \end{array} \right\}.$$

Necessary and sufficient conditions for local invertability of Lipschitz functions can be given in terms of Thibault derivatives.

Theorem 16 (Kummer's inverse function theorem [81, 86]). *Let $F : \mathbb{R}^n \longrightarrow \mathbb{R}^n$ be Lipschitz near $\bar{x}$. Then the following statements are equivalent:*

(i)   *F has the locally unique Lipschitz inverse function F^{-1}.*

(ii)   *There exists $c > 0$ such that*

$$\|F(x) - F(x')\| \geq c\|x - x'\| \text{ for all } x, x' \text{ with } \|\bar{x} - x\| \leq c, \|\bar{x} - x'\| \leq c.$$

(iii)   *$TF(\bar{x})$ is injective (i.e., $0 \notin TF(\bar{x})(u)$ for all $u \neq 0$).*

Remark 7. Note that the injectivity of $TF(\bar{x})$ in Theorem 16 is in general weaker than Clarke's requirement that all matrices in $\partial F(\bar{x})$ are nonsingular. In fact, there exists a Lipschitz homeomorphism F of $\mathbb{R}^2$ such that $\partial F(\bar{x})$ contains the zero matrix (see Example BE.3 in [81]).

Remark 8. We point out that (iii) from Theorem 16 implies the existence of the unique Lipschitz inverse of F w.r.t. Lipschitz perturbations of F performed locally. This means that there exists an $\mathbb{R}^n$-neighborhood U of $\bar{x}$ and a neighborhood V of F in the space $C^{0,1}(U, \mathbb{R}^n)$ of Lipschitz functions such that, for all $\widehat{F} \in V$ and $\widehat{x} \in U$, $\widehat{F}$ has the locally unique Lipschitz inverse function $\widehat{F}^{-1}$ around $\widehat{x}$. Note that we equip $F \in C^{0,1}(U, \mathbb{R}^n)$ with the norm

$$|F| := \max \left\{ \sup_{x \in U} \|F(x)\| + \mathrm{Lip}(F, U) \right\},$$

where

$$\text{Lip}(F,U) := \inf \left\{ r > 0 \,\middle|\, \|F(x) - F(x')\| \le r\|x - x'\| \text{ for all } x, x' \in U \right\}.$$

For details, we refer the reader to Theorem 5.14 and Corollary 4.4 in [81].

Theorem 17 (Kummer's implicit function theorem, [81, 85]). *Let* $G : \mathbb{R}^{n-k} \times \mathbb{R}^k \longrightarrow \mathbb{R}^k$ *be Lipschitz near* $(\bar{y}, \bar{z}) \in \mathbb{R}^{n-k} \times \mathbb{R}^k$ *with* $G(\bar{y}, \bar{z}) = 0$. *Then, the following statements are equivalent:*

(i)   *There exist* $\mathbb{R}^{n-k}$*-neighborhoods* Y *of* $\bar{y}$ *and* W *of* 0, *an* $\mathbb{R}^k$*-neighborhood* Z *of* $\bar{z}$, *and a Lipschitz function* $\zeta : Y \times W \longrightarrow Z$ *such that* $\zeta(\bar{y}, 0) = \bar{z}$ *and for every* $(y, z, w) \in Y \times Z \times W$ *it holds that*

$$G(y, z) = w \text{ if and only if } z = \zeta(y, w).$$

(ii) $0 \notin TG(\bar{y}, \bar{z})(0, u)$ *for all* $u \ne 0$.

Remark 9. We point out that Theorem 17 gives a necessary and sufficient condition for the existence of implicit functions. Recall that Clarke's IFT (see Theorem 14) gives only a sufficient condition for that fact. Moreover, it is important to note that in Theorem 17 the implicit function ζ depends Lipschitz also on the right-hand-side perturbations w. This issue was used extensively in the proof of Theorem 15.

Now, we turn our attention to the case of min-functions. Let a basis decomposition of $\mathbb{R}^n = \mathbb{R}^{n-k} \times \mathbb{R}^k$ be fixed. It turns out that the assumptions of Clarke's and Kummer's implicit function theorems coincide. Moreover, they are also equivalent with the SMFC w.r.t. the subspace $E := \{0_{n-k}\} \times \mathbb{R}^k$ (B. Kummer and O. Stein, personal communication).

Lemma 4. *Setting* $g_i := \min\{F_{1,i}, F_{2,i}\}$, $i = 1, \dots, k$, *we define*

$$G : \begin{cases} \mathbb{R}^{n-k} \times \mathbb{R}^k \longrightarrow \mathbb{R}^k, \\ (y, z) \mapsto (g_1(y, z), \dots, g_k(y, z)). \end{cases}$$

Then, the following conditions are equivalent for $\bar{x} = (\bar{y}, \bar{z})$*:*

(i) $\pi_z \partial G(\bar{y}, \bar{z})$ *is of maximal rank, meaning*

$$\pi_z \partial G(\bar{y}, \bar{z}) := \{M \in \mathbb{R}^{k \times k} \,|\, \text{there exists } N \in \mathbb{R}^{k \times n} \text{ with } [N, M] \in \partial G(\bar{y}, \bar{z})\}$$

contains merely nonsingular matrices.

(ii)   *All matrices in* $\partial_z g_1(\bar{x}) \times \partial_z g_2(\bar{x}) \times \dots \times \partial_z g_k(\bar{x})$ *are nonsingular.*

(iii) $0 \notin TG(\bar{y}, \bar{z})(0, u)$ *for all* $u \ne 0$.

Proof. "**(i)** $\Longrightarrow$ **(iii)**": Due to (i), we may apply Clarke's implicit function theorem. Hence, the implicit function $\zeta(y)$ exists. It is not hard to see that ζ depends uniquely and Lipschitz on the w-values of G. Hence, we obtain in fact $\zeta(y, w)$. Applying Kummer's implicit function theorem, we get (iii).

"(ii) $\Longrightarrow$ (i)": In general, it holds (see, e.g., [23]) that

$$\pi_z \partial G(\bar{y},\bar{z}) \subset \partial_z g_1(\bar{x}) \times \partial_z g_2(\bar{x}) \times \ldots \times \partial_z g_k(\bar{x}).$$

This inclusion shows the assertion.

"(iii) $\Longrightarrow$ (ii)": Let $0 \notin TG(\bar{y},\bar{z})(0,u)$ for all $u \neq 0$. For $q \in \mathbb{R}^k$, we set

$$q^+ := (q_1^+,\ldots,q_k^+), \text{ where } q_i^+ := \max\{q,0\}, i = 1,\ldots,k,$$

$$q^- := (q_1^-,\ldots,q_k^-), \text{ where } q_i^- := \min\{q,0\}, i = 1,\ldots,k.$$

We define the mapping $\widehat{G} : \mathbb{R}^{n-k} \times \mathbb{R}^k \times \mathbb{R}^k \longrightarrow \mathbb{R}^{2k}$ as

$$\widehat{G}(y,z,q) = \begin{pmatrix} F_1(y,z) - q^+ \\ -F_2(y,z) - q^- \end{pmatrix}.$$

The zeros of G and $\widehat{G}$ correspond as follows. If $G(x) = 0$, then $\widehat{G}(x,q) = 0$ with $q := F_1 - F_2$. If $\widehat{G}(x,q) = 0$, then $G(x) = 0$.

Setting $\bar{q} = F_1(\bar{y},\bar{z}) - F_2(\bar{y},\bar{z})$, we claim that $0 \notin T\widehat{G}(\bar{x},\bar{q})(0,u,p)$ for all $(u,p) \neq 0$. In fact, from Kummer's IFT, the latter is equivalent to the existence of Lipschitz implicit functions $\zeta(y,w_1,w_2)$ and $q(y,w_1,w_2)$ for the system

$$F_1(y,z) - q^+ = w_1, \ -F_2(y,z) - q^- = w_2. \tag{2.10}$$

The system (2.10) can be equivalently written as

$$F_1(y,z) - w_1 - q^+ = 0, \ -F_2(y,z) - w_2 - q^- = 0.$$

Hence, we need to find the implicit function $\zeta(y,w_1,w_2)$ for

$$\min\{F_{1,i}(y,z) - w_{1,i}, F_{2,i}(y,z) + w_{2,i}\} = 0, i = 1,\ldots,k \tag{2.11}$$

and afterwards to set

$$q(y,w_1,w_2) := F_1(y,\zeta(y,w_1,w_2)) - w_1 - F_2(y,\zeta(y,w_1,w_2)) - w_2.$$

Note that (2.11) is a Lipschitz perturbed version of $G(x) = 0$. Remark 8 and (iii) then justify the application of Kummer's IFT for the perturbed system (2.11). Hence, $0 \notin T\widehat{G}(\bar{x},\bar{q})(0,u,p)$ for all $(u,p) \neq 0$.

Now, we compute $T\widehat{G}(\bar{x},\bar{q})(0,u,p)$ using results from [81] on the so-called Kojima functions. For that, we set

$$N(q) := (1,q^+,q^-) \text{ and } M(x) := \begin{pmatrix} F_1(x) & F_2(x) \\ -I_k & 0 \\ 0 & -I_k \end{pmatrix},$$

where I_k is the $k \times k$ identity matrix. It holds that

$$\widehat{F}(x,q) = N(q) \cdot M(x).$$

Applying the product rule (see Theorem 7.5 in [81]), we get:

$$T\widehat{F}(\bar{x},\bar{q})(0,u,p) = N(\bar{q})TM(\bar{x})(0,u) + TN(\bar{q})(p)M(\bar{x}).$$

We compute $TM(\bar{x})$ and $TN(\bar{q})$ (see Lemma 7.3 in [81]) as

$$TM(\bar{x})(0,u) = \begin{pmatrix} D_zF_1(\bar{x})u & D_zF_2(\bar{x})u \\ 0 & 0 \\ 0 & 0 \end{pmatrix},$$

$$TN(\bar{q})(p) = \{(0,\lambda,p-\lambda)\,|\,\lambda_i = r_ip_i, \, r_i \in \mathfrak{R}(\bar{q}), \, i = 1,\ldots,k\}, \text{ where}$$

$$\mathfrak{R}(\bar{q}) := \{r \in [0,1]^m \,|\, r_i = 1 \text{ if } \bar{q}_i > 0, \, r_i = 0 \text{ if } \bar{q}_i < 0\}.$$

Combining the above

$$T\widehat{G}(\bar{x},\bar{q})(0,u,p) = \left\{ \begin{pmatrix} D_zF_{1,i}(\bar{x})u - r_ip_i, \, i = 1,\ldots,k \\ -D_zF_{2,i}(\bar{x})u - (1-r_i)p_i, \, i = 1,\ldots,k \end{pmatrix} \,\middle|\, r \in \mathfrak{R}(\bar{q}) \right\}.$$

Next, let (ii) beassumed to fail. We show that this contradicts the fact that

$$0 \notin T\widehat{G}(\bar{x},\bar{q})(0,u,p) \text{ for all } (u,p) \neq 0.$$

Indeed, if (ii) does not hold, we obtain $u \in \mathbb{R}^k$, $u \neq 0$, and $r \in [0,1]^k$ such that

$$[(1-r_i)D_zF_{1,i}(\bar{x}) + r_iD_zF_{2,i}(\bar{x})]u = 0, \, i = 1,\ldots,k. \tag{2.12}$$

Note that $r \in \mathfrak{R}(\bar{q})$ due to the definition of Clarke's subdifferentials.

Case $r_i \neq 0$. Then, we set $p_i := \dfrac{1}{r_i}D_zF_{1,i}u$ and obtain

$$D_zF_{1,i}(\bar{x})u - r_ip_i = 0.$$

Furthermore, from (2.12) we get,

$$r_i[-D_zF_{2,i}(\bar{x})u - (1-r_i)p_i] = r_i\left[-D_zF_{2,i}(\bar{x})u - (1-r_i)\frac{1}{r_i}D_zF_{1,i}(\bar{x})u\right]$$

$$= -r_iD_zF_{2,i}(\bar{x})u - (1-r_i)D_zF_{1,i}(\bar{x})u = 0.$$

Hence, $-D_zF_{2,i}(\bar{x})u - (1-r_i)p_i = 0$.

Case $r_i = 0$. Then, we set $p_i := -D_zF_{2,i}u$ and obtain from (2.12)

$$D_zF_{1,i}(\bar{x})u - r_ip_i = 0.$$

Moreover,

$$-D_zF_{2,i}(\bar{x})u - (1-r_i)p_i = -D_zF_{2,i}(\bar{x})u + D_zF_{2,i}u = 0.$$

Thus, we see that, for $(u, p) \neq 0$ defined as above, it holds that

$$0 \in T\widehat{G}(\bar{x}, \bar{q})(0, u, p). \qquad \square$$

From Lemma 4, we deduce the following result.

Theorem 18 (SMFC and Kummer's implicit function theorem). *The SMFC holds if and only if Kummer's implicit function theorem is applicable w.r.t. some basis decomposition of* $\mathbb{R}^n$.

Proof. The equivalence of (ii) and (iii) from Lemma 4 immediately inplies the result. In fact, we only need to use the chain rule from [81],

$$T(G \circ A)(x)(u) = TG(Ax)(Au),$$

where A is a nonsingular $(n \times n)$ matrix. See also the characterization of the SMFC in terms of Clarke's subdifferentials in Lemma 1. $\square$

Theorem 18 shows that the remaining difficulty concerning topological stability of the MPCC feasible set lies in the conjectured equivalence between the MFC and SMFC rather than in an application of different implicit function theorems.

2.3 Critical point theory

We study the behavior of the topological properties of lower-level sets

$$M^a := \{x \in M \mid f(x) \le a\}$$

as the level $a \in \mathbb{R}$ varies. It turns out that the concept of C-stationarity is an adequate stationarity concept. In fact, we present two basic theorems from Morse theory (see [63, 93]). First, we show that, for $a < b$, the set M^a is a strong deformation retract of M^b if the (compact) set

$$M_a^b := \{x \in M \mid a \le f(x) \le b\}$$

does not contain C-stationary points (see Theorem 20(a)). Second, if M_a^b contains exactly one (nondegenerate) C-stationary point, then M^b is shown to be homotopy-equivalent to M^a with a q-cell attached (see Theorem 20(b)). Here, the dimension q is the so-called C-index. It depends on both the restricted Hessian of the Lagrangian and the Lagrange multipliers related to biactive complementarity constraints. The latter fact is the main difference with respect to the well-known case where a feasible set is described only by equality and finitely many inequality constraints (see [63] and Section 1.4).

We would like to refer to some related papers. In [106], the concept of a non-degenerate feasible point for the MPCC is introduced. Some genericity results are obtained. In [99], the concepts of a nondegenerate C-stationary point and its stationary C-index are introduced for quadratic programs with complementarity constraints

(QPCCs). The generic structure of the C-stationary point set for nonparametric and one-parametric QPCCs is discussed, and some homotopy methods for QPCCs are developed. We refer the reader to [69] for details.

Notation and Auxiliary Results

Given $\bar{x} \in M$, we define the following index sets:

$$J_0(\bar{x}) := \{ j \in J \mid g_j(\bar{x}) = 0 \},$$

$$\alpha(\bar{x}) := \{ m \in \{1, \ldots k\} \mid F_{1,m}(\bar{x}) = 0, F_{2,m}(\bar{x}) > 0 \},$$

$$\beta(\bar{x}) := \{ m \in \{1, \ldots k\} \mid F_{1,m}(\bar{x}) = 0, F_{2,m}(\bar{x}) = 0 \},$$

$$\gamma(\bar{x}) := \{ m \in \{1, \ldots k\} \mid F_{1,m}(\bar{x}) > 0, F_{2,m}(\bar{x}) = 0 \}.$$

We call $J_0(\bar{x})$ the active inequality index set and $\beta(\bar{x})$ the biactive index set at $\bar{x}$.

Without loss of generality, we assume that at the particular point of interest $\bar{x} \in M$ it holds that

$$J_0(\bar{x}) = \{1, \ldots, |J_0(\bar{x})|\}, \ \alpha(\bar{x}) = \{1, \ldots, |\alpha(\bar{x})|\},$$

$$\gamma(\bar{x}) = \{|\alpha(\bar{x})| + 1, \ldots, |\alpha(\bar{x})| + |\gamma(\bar{x})|\}.$$

We put $s := |I| + |\alpha(\bar{x})| + |\gamma(\bar{x})|$, $q := s + |J_0(\bar{x})|$, $p := n - q - 2|\beta(\bar{x})|$.

Furthermore, we recall the well-known linear independence constraint qualification (LICQ) for the MPCC (e.g. [105]), which is said to hold at $\bar{x} \in M$ if the set of vectors

$$\{ D^T h_i(\bar{x}), i \in I, D^T F_{1,m_\alpha}(\bar{x}), m_\alpha \in \alpha(\bar{x}), D^T F_{2,m_\gamma}(\bar{x}), m_\gamma \in \gamma(\bar{x}),$$
$$D^T g_j(\bar{x}), j \in J_0(\bar{x}), D^T F_{1,m_\beta}(\bar{x}), D^T F_{2,m_\beta}(\bar{x}), m_\beta \in \beta(\bar{x}) \}$$

is linearly independent.

Definition 10 (C-stationary point [22, 105]). A point $\bar{x} \in M$ is called Clarke stationary (C-stationary) for the MPCC if there exist real numbers $\bar{\lambda}_i$, $i \in I$, $\bar{\rho}_{m_\alpha}$, $m_\alpha \in \alpha(\bar{x})$, $\bar{\vartheta}_{m_\gamma}$, $m_\gamma \in \gamma(\bar{x})$, $\bar{\mu}_j$, $j \in J_0(\bar{x})$, $\bar{\sigma}_{1,m_\beta}$, $\bar{\sigma}_{2,m_\beta}$, $m_\beta \in \beta(\bar{x})$ (Lagrange multipliers) such that

$$Df(\bar{x}) = \sum_{i \in I} \bar{\lambda}_i Dh_i(\bar{x}) + \sum_{m_\alpha \in \alpha(\bar{x})} \bar{\rho}_{m_\alpha} DF_{1,m_\alpha}(\bar{x}) + \sum_{m_\gamma \in \gamma(\bar{x})} \bar{\vartheta}_{m_\gamma} DF_{2,m_\gamma}(\bar{x})$$

$$+ \sum_{j \in J_0(\bar{x})} \bar{\mu}_j Dg_j(\bar{x}) + \sum_{m_\beta \in \beta(\bar{x})} \left(\bar{\sigma}_{1,m_\beta} DF_{1,m_\beta}(\bar{x}) + \bar{\sigma}_{2,m_\beta} DF_{2,m_\beta}(\bar{x}) \right), \quad (2.13)$$

$$\bar{\mu}_j \geq 0 \text{ for all } j \in J_0(\bar{x}), \quad (2.14)$$

$$\bar{\sigma}_{1,m_\beta} \cdot \bar{\sigma}_{2,m_\beta} \geq 0 \text{ for all } m_\beta \in \beta(\bar{x}). \quad (2.15)$$

In the case where the LICQ holds at $\bar{x} \in M$, the Lagrange multipliers in (2.13) are uniquely determined.

Given a C-stationary point $\bar{x} \in M$ for the MPCC, we set

$$M(\bar{x}) := \{x \in \mathbb{R}^n \mid h_i(x) = 0, i \in I, F_{1,m_\alpha}(x) = 0, m_\alpha \in \alpha(\bar{x}),$$
$$F_{2,m_\gamma}(x) = 0, m_\gamma \in \gamma(\bar{x}), g_j(x) = 0, j \in J_0(\bar{x}),$$
$$F_{1,m_\beta}(x) = 0, F_{2,m_\beta}(x) = 0, m_\beta \in \beta(\bar{x})\}.$$

Obviously, $M(\bar{x}) \subset M$ and, in the case where the LICQ holds at $\bar{x}$, $M(\bar{x})$ is locally a p-dimensional C^2-manifold.

Definition 11 (Nondegenerate C-stationary point [99, 106]). A C-stationary point $\bar{x} \in M$ with Lagrange multipliers as in Definition 10 is called nondegenerate if the following conditions are satisfied:

ND1:   LICQ holds at $\bar{x}$.

ND2: $\bar{\mu}_j > 0$ for all $j \in J_0(\bar{x})$.

ND3: $D^2L(\bar{x})|_{T_{\bar{x}}M(\bar{x})}$ is nonsingular.

ND4: $\bar{\sigma}_{1,m_\beta} \cdot \bar{\sigma}_{2,m_\beta} > 0$ for all $m_\beta \in \beta(\bar{x})$.

Here, the matrix D^2L stands for the Hessian of the Lagrange function L,

$$L(x) := f(x) - \sum_{i \in I} \bar{\lambda}_i h_i(x) - \sum_{m_\alpha \in \alpha(\bar{x})} \bar{\rho}_{m_\alpha} F_{1,m_\alpha}(x) - \sum_{m_\gamma \in \gamma(\bar{x})} \bar{\vartheta}_{m_\gamma} F_{2,m_\gamma}(x)$$

$$- \sum_{j \in J_0(\bar{x})} \bar{\mu}_j g_j(x) - \sum_{m_\beta \in \beta(\bar{x})} \left(\bar{\sigma}_{1,m_\beta} F_{1,m_\beta}(x) + \bar{\sigma}_{2,m_\beta} F_{2,m_\beta}(x) \right), \qquad (2.16)$$

and $T_{\bar{x}}M(\bar{x})$ denotes the tangent space of $M(\bar{x})$ at $\bar{x}$,

$$T_{\bar{x}}M(\bar{x}) := \{\xi \in \mathbb{R}^n \mid Dh_i(\bar{x})\xi = 0, i \in I,$$
$$DF_{1,m_\alpha}(\bar{x})\xi = 0, m_\alpha \in \alpha(\bar{x}),$$
$$DF_{2,m_\gamma}(\bar{x})\xi = 0, m_\gamma \in \gamma(\bar{x}),$$
$$Dg_j(\bar{x})\xi = 0, j \in J_0(\bar{x}),$$
$$DF_{1,m_\beta}(\bar{x})\xi = 0, DF_{2,m_\beta}(\bar{x})\xi = 0, m_\beta \in \beta(\bar{x})\}.$$

Condition ND3 means that the matrix $V^T D^2 L(\bar{x}) V$ is nonsingular, where V is some matrix whose columns form a basis for the tangent space $T_{\bar{x}}M(\bar{x})$.

Definition 12 (C-index [99]). Let $\bar{x} \in M$ be a nondegenerate C-stationary point with Lagrange multipliers as in Definition 11. The number of negative/positive eigenvalues of $D^2L(\bar{x})|_{T_{\bar{x}}M(\bar{x})}$ is called the quadratic index (QI)/quadratic coindex (QCI) of $\bar{x}$. The number of pairs $(\bar{\sigma}_{1,m_\beta}, \bar{\sigma}_{2,m_\beta})$, $m_\beta \in \beta(\bar{x})$ with both $\bar{\sigma}_{1,m_\beta}$ and $\bar{\sigma}_{2,m_\beta}$ negative/positive is called the biactive index (BI)/biactive coindex (BCI) of $\bar{x}$. The number $(QI + BI)/(QCI + BCI)$ is called the Clarke index (C-index)/Clarke coindex (C-coindex) of $\bar{x}$.

Note that, in the absence of complementarity constraints, the C-index has only the QI part and coincides with the well-known quadratic index of a nondegenerate Karush-Kuhn-Tucker point in nonlinear programming or, equivalently, with the Morse index (see [63, 83, 93] and Section 1.4).

The following proposition uses the C-index for the characterization of a local minimizer. Its proof is omitted since it can be easily seen (see also [99, 105]).

Proposition 5. (i)   Assume that $\bar{x}$ is a local minimizer for the MPCC and that the LICQ holds at $\bar{x}$. Then, $\bar{x}$ is a C-stationary point for the MPCC.

(ii)   Let $\bar{x}$ be a nondegenerate C-stationary point for the MPCC. Then, $\bar{x}$ is a local minimizer for the MPCC if and only if its C-index is equal to zero.

The next proposition concerning genericity results for the LICQ and for nondegeneracy of C-stationary points mainly follows from [63]. It was shown in [106] and for the special case of the QPCC in [99].

Proposition 6 (Genericity and Stability [99, 106]).

(i)   Let $\mathscr{F}$ denote the subset of

$$C^2(\mathbb{R}^n, \mathbb{R}^{|I|}) \times C^2(\mathbb{R}^n, \mathbb{R}^{|J|}) \times C^2(\mathbb{R}^n, \mathbb{R}^k) \times C^2(\mathbb{R}^n, \mathbb{R}^k)$$

consisting of those (h, g, F_1, F_2) for which the LICQ holds at all points $x \in M[h, g, F_1, F_2]$. Then, $\mathscr{F}$ is C_s^2-open and -dense.

(ii)   Let $\mathscr{D}$ denote the subset of

$$C^2(\mathbb{R}^n, \mathbb{R}) \times C^2(\mathbb{R}^n, \mathbb{R}^{|I|}) \times C^2(\mathbb{R}^n, \mathbb{R}^{|J|}) \times C^2(\mathbb{R}^n, \mathbb{R}^k) \times C^2(\mathbb{R}^n, \mathbb{R}^k)$$

consisting of those (f, h, g, F_1, F_2) for which each C-stationary point of the MPCC with data functions (f, h, g, F_1, F_2) is nondegenerate. Then, $\mathscr{D}$ is C_s^2-open and -dense.

Morse lemma for the MPCC

For the proof of deformation and cell-attachment results, we locally describe the MPCC feasible set under the LICQ (see Lemma 5). Moreover, an equivariant Morse lemma for the MPCC is derived in order to obtain suitable normal forms for the objective function at C-stationary points (see Theorem 19).

Definition 13. The feasible set M admits a local C^r-coordinate system of $\mathbb{R}^n$ ($r \geq 1$) at $\bar{x}$ by means of a C^r-diffeomorphism $\Phi : U \longrightarrow V$ with open $\mathbb{R}^n$-neighborhoods U and V of $\bar{x}$ and 0, respectively, if it holds that

(i) $\Phi(\bar{x}) = 0$,

(ii) $\Phi(M \cap U) = \left(\{0_s\} \times \mathbb{H}^{|J_0(\bar{x})|} \times \left(\partial \mathbb{H}^2 \right)^{|\beta(\bar{x})|} \times \mathbb{R}^p \right) \cap V$.

Lemma 5 (see also [106]). Suppose that the LICQ holds at $\bar{x} \in M$. Then M admits a local C^2-coordinate system of $\mathbb{R}^n$ at $\bar{x}$.

Proof. Choose vectors $\xi_l \in \mathbb{R}^n$, $l = 1, \ldots, p$, which together with the vectors

$$\{D^T h_i(\bar{x}), i \in I,, D^T F_{1,m_\alpha}(\bar{x}), m_\alpha \in \alpha(\bar{x}), D^T F_{2,m_\gamma}(\bar{x}), m_\gamma \in \gamma(\bar{x}),$$
$$D^T g_j(\bar{x}), j \in J_0(\bar{x}), D^T F_{1,m_\beta}(\bar{x}), D^T F_{2,m_\beta}(\bar{x}), m_\beta \in \beta(\bar{x})\},$$

form a basis for $\mathbb{R}^n$. Next, we put

$$
\left.
\begin{aligned}
y_i &:= h_i(x), \, i \in I, \\
y_{|I|+m_\alpha} &:= F_{1,m_\alpha}(x), \, m_\alpha \in \alpha(\bar{x}), \\
y_{|I|+m_\gamma} &:= F_{2,m_\gamma}(x), \, m_\gamma \in \gamma(\bar{x}), \\
y_{s+j} &:= g_j(x), \, j \in J_0(\bar{x}), \\
y_{s+|J_0(\bar{x})|+2m_\beta-1} &:= F_{1,m_\beta}(x), \\
y_{s+|J_0(\bar{x})|+2m_\beta} &:= F_{2,m_\beta}(x), \, m_\beta = 1, \ldots, |\beta(\bar{x})|, \\
y_{n-p+l} &:= \xi_l^T(x - \bar{x}), \, l = 1, \ldots, p,
\end{aligned}
\right\}
\tag{2.17}
$$

or, for short,

$$y = \Phi(x). \tag{2.18}$$

Note that $\Phi \in C^2(\mathbb{R}^n, \mathbb{R}^n)$, $\Phi(\bar{x}) = 0$, and the Jacobian matrix $D\Phi(\bar{x})$ is nonsingular (by virtue of the LICQ and the choice of ξ_l, $l = 1, \ldots, p$). By means of the implicit function theorem, there exist open neighborhoods U of $\bar{x}$ and V of 0 such that $\Phi : U \longrightarrow V$ is a C^2-diffeomorphism. By shrinking U if necessary, we can guarantee that $J_0(x) \subset J_0(\bar{x})$ and $\beta(x) \subset \beta(\bar{x})$ for all $x \in M \cap U$. Thus, property (ii) in Definition 13 follows directly from the definition of Φ. $\square$

Definition 14. We will refer to the C^2-diffeomorphism Φ defined by (2.17) and (2.18) as a standard diffeomorphism.

Remark 10. From the proof of Lemma 5, it follows that the Lagrange multipliers at a nondegenerate C-stationary point are the corresponding partial derivatives of the objective function in new coordinates given by the standard diffeomorphism (see Lemma 2.2.1 of [65]). Moreover, the Hessian with respect to the last p coordinates corresponds to the restriction of the Lagrange function's Hessian on the respective tangent space (see Lemma 2.2.10 of [65]).

Theorem 19 (Morse lemma for MPCC). *Suppose that $\bar{x}$ is a nondegenerate C-stationary point for the MPCC with quadratic index QI, biactive index BI, and C-index = QI + BI. Then, there exists a local C^1-coordinate system $\Psi : U \longrightarrow V$ of $\mathbb{R}^n$ around $\bar{x}$ (according to Definition 13) such that*

$$f \circ \Psi^{-1}(0_s, y_{s+1}, \ldots, y_n) =$$

$$f(\bar{x}) + \sum_{i=1}^{|J_0(\bar{x})|} y_{i+s} + \sum_{j=1}^{|\beta(\bar{x})|} \pm \left(y_{2j+q-1} + y_{2j+q}\right) + \sum_{k=1}^{p} \pm y_{k+n-p}^2, \tag{2.19}$$

where $y \in \{0_s\} \times \mathbb{H}^{|J_0(\bar{x})|} \times \left(\partial\mathbb{H}^2\right)^{|\beta(\bar{x})|} \times \mathbb{R}^p$. Moreover, in (2.19) there are exactly BI negative linear pairs and QI negative squares.

Proof. Without loss of generality, we may assume $f(\bar{x}) = 0$. Let $\Phi : U \longrightarrow V$ be a standard diffeomorphism according to Definition 14. We put $\bar{f} := f \circ \Phi^{-1}$ on the set $\left(\{0_s\} \times \mathbb{H}^{|J_0(\bar{x})|} \times \left(\partial \mathbb{H}^2\right)^{|\beta(\bar{x})|} \times \mathbb{R}^p\right) \cap V$. From now on, we may assume $s = 0$. In view of Remark 10 we have at the origin

(i) $\dfrac{\partial \bar{f}}{\partial y_i} > 0$, $i \in J_0(\bar{x})$,

(ii) $\dfrac{\partial \bar{f}}{\partial y_{2j+q-1}} \cdot \dfrac{\partial \bar{f}}{\partial y_{2j+q}} > 0$, $j = 1, \ldots |\beta(\bar{x})|$,

(iii) $\dfrac{\partial \bar{f}}{\partial y_{2j+q-1}} < 0$ for exactly BI indices $j \in \{1, \ldots |\beta(\bar{x})|\}$,

(iv) $\dfrac{\partial \bar{f}}{\partial y_{k+n-p}} = 0$, $k = 1, \ldots, p$ and $\left(\dfrac{\partial^2 \bar{f}}{\partial y_{k_1+n-p} \partial y_{k_2+n-p}}\right)_{1 \leq k_1, k_2 \leq p}$ is a nonsingular matrix with QI negative eigenvalues.

From now on, we denote $\bar{f}$ by f. Under the following coordinate transformations, the set $\mathbb{H}^{|J_0(\bar{x})|} \times \left(\partial \mathbb{H}^2\right)^{|\beta(\bar{x})|} \times \mathbb{R}^p$ will be transformed in itself (equivariant). As an abbreviation, we put $y = (Y_{n-p}, Y^p)$, where $Y_{n-p} = (y_1, \ldots, y_{n-p})$ and $Y^p = (y_{n-p+1}, \ldots, y_n)$. We write

$$f(Y_{n-p}, Y^p) = f(0, Y^p) + \int_0^1 \frac{d}{dt} f(tY_{n-p}, Y^p) dt = f(0, Y^p) + \sum_{i=1}^{n-p} y_i d_i(y),$$

where $d_i \in C^1$, $i = 1, \ldots, n - p$.

In view of (iv), we may apply the Morse lemma on the C^2-function $f(0, Y^p)$ (see Theorem 2.8.2 of [63]) without affecting the coordinates Y_{n-p}. The corresponding coordinate transformation is of class C^1. Denoting the transformed functions f, d_j again by f, d_j, we obtain

$$f(y) = \sum_{i=1}^{n-p} y_i d_i(y) + \sum_{k=1}^{p} \pm y_{k+n-p}^2.$$

Note that $d_i(0) = \dfrac{\partial f}{\partial y_i}(0)$, $i = 1, \ldots, n - p$. Recalling (i)–(iii), we have

$$y_i |d_i(y)|, i = 1, \ldots, n - p, \quad y_j, j = n - p + 1, \ldots, n \qquad (2.20)$$

as new local C^1-coordinates. Denoting the transformed function f again by f and recalling the signs in (i)–(iii), we obtain (2.19). Here, the coordinate transformation Ψ is understood as the composite of all previous ones. $\square$

Theorem 19 allows us to provide two other local representations (normal forms) of the objective function on the MPCC feasible set with respect to Lipschitz and Hölder coordinate systems.

Recall that the set $\partial \mathbb{H}^2$ represents the complementarity relations

$$u \geq 0, \, v \geq 0, \, u \cdot v = 0.$$

Define the mapping $\varphi : \partial \mathbb{H}^2 \longrightarrow \mathbb{R}^1 \times 0_1$ as

$$\varphi(u,0) := (u,0), \; \varphi(0,v) := (-v,0). \tag{2.21}$$

By coordinatewise extension of φ on $\left(\partial \mathbb{H}^2\right)^{|\beta(\bar{x})|}$ and leaving the other coordinates invariant, (2.21) induces the Lipschitz coordinate transformation Φ,

$$\Phi : \{0_s\} \times \mathbb{H}^{|J_0(\bar{x})|} \times \left(\partial \mathbb{H}^2\right)^{|\beta(\bar{x})|} \times \mathbb{R}^p \longrightarrow \mathbb{H}^{|J_0(\bar{x})|} \times \mathbb{R}^{|\beta(\bar{x})|} \times \mathbb{R}^p \tag{2.22}$$

On the right-hand side of (2.22), the zeros $\{0_s\}$ and $\{0_1\}$ ($|\beta(\bar{x})|$ times) are deleted. The proof of the following corollary is now straightforward.

Corollary 3 (Normal forms in Lipschitz coordinates). *Let f have the normal form as in (2.19), and let Φ be the Lipschitz coordinate transformation (2.22). Then, we have*

$$f \circ \Phi^{-1}(y) = f(\bar{x}) + \sum_{i=1}^{|J_0(\bar{x})|} y_i + \sum_{j=|J_0(\bar{x})|+1}^{J_0(\bar{x})+|\beta(\bar{x})|} \pm |y_j| + \sum_{k=J_0(\bar{x})+|\beta(\bar{x})|+1}^{n-|\beta(\bar{x})|+s} \pm y_{k+n-p}^2. \tag{2.23}$$

In (2.23), there are exactly BI negative absolute value terms and QI negative squares.

On $\mathbb{R}^1$, we introduce the transformation ψ:

$$\psi(y) := \mathrm{sgn}(y)\sqrt{|y|}. \tag{2.24}$$

Note that the function $\pm|y|$ transforms into $\pm y^2$ under ψ^{-1}. By coordinatewise extension of ψ on $\mathbb{R}^{|\beta(\bar{x})|}$ and leaving the other coordinates invariant, (2.24) induces the Hölder coordinate transformation Ψ,

$$\Psi : \mathbb{H}^{|J_0(\bar{x})|} \times \mathbb{R}^{|\beta(\bar{x})|} \times \mathbb{R}^p \longrightarrow \mathbb{H}^{|J_0(\bar{x})|} \times \mathbb{R}^{|\beta(\bar{x})|} \times \mathbb{R}^p. \tag{2.25}$$

The proof of the following corollary is again straightforward.

Corollary 4 (Normal forms in Hölder coordinates). *Let f have the normal form as in (2.23), and let Ψ be the Hölder coordinate transformation (2.25). Then, we have*

$$f \circ \Psi^{-1}(y) = f(\bar{x}) + \sum_{i=1}^{|J_0(\bar{x})|} y_i + \sum_{j=|J_0(\bar{x})|+1}^{n-|\beta(\bar{x})|+s} \pm y_j^2. \tag{2.26}$$

The number of negative squares in (2.26) equals the C-index BI+QI.

Deformation and Cell Attachment

We state and prove the main deformation and cell-attachment theorems for the MPCC. Recall that for $a, b \in \mathbb{R}$, $a < b$ the sets M^a and M_a^b are defined as

$$M^a := \{x \in M \mid f(x) \leq a\}$$

and

$$M_a^b := \{x \in M \mid a \leq f(x) \leq b\}.$$

Theorem 20. *Let M_a^b be compact, and suppose that the LICQ is satisfied at all points $x \in M_a^b$.*

(a) **(Deformation theorem)** *If M_a^b does not contain any C-stationary point for the MPCC, then M^a is a strong deformation retract of M^b.*

(b) **(Cell-attachment theorem)** *If M_a^b contains exactly one C-stationary point for the MPCC, say $\bar{x}$, and if $a < f(\bar{x}) < b$ and the C-index of $\bar{x}$ is equal to q, then M^b is homotopy-equivalent to M^a with a q-cell attached.*

Proof. (a) Due to the LICQ at all $x \in M_a^b$, there exist real numbers $\lambda_i(x), i \in I$, $\rho_{m_\alpha}(x), m_\alpha \in \alpha(x), \vartheta_{m_\gamma}(x), m_\gamma \in \gamma(x), \mu_j(x), j \in J_0(x), \sigma_{1,m_\beta}(x), \sigma_{2,m_\beta}(x), m_\beta \in \beta(x), \nu_l(x), l = 1, \ldots, p$ such that

$$Df(x) = \sum_{i \in I} \lambda_i(x) Dh_i(x) + \sum_{m_\alpha \in \alpha(x)} \rho_{m_\alpha}(x) DF_{1,m_\alpha}(x)$$

$$+ \sum_{m_\gamma \in \gamma(x)} \vartheta_{m_\gamma}(x) DF_{2,m_\gamma}(x) + \sum_{j \in J_0(x)} \mu_j(x) Dg_j(x)$$

$$+ \sum_{m_\beta \in \beta(x)} \left(\sigma_{1,m_\beta}(x) DF_{1,m_\beta}(x) + \sigma_{2,m_\beta}(x) DF_{2,m_\beta}(x) \right) + \sum_{l=1}^{p} \nu_l(x) \xi_l,$$

where vectors $\xi_l, l = 1, \ldots, p$ are chosen as in Lemma 5. We set:

$$A := \{x \in M_a^b \mid \text{there exists } l \in \{1, \ldots, p\} \text{ with } \nu_l(x) \neq 0\},$$

$$B := \{x \in M_a^b \mid \text{there exists } j \in J_0(x) \text{ with } \mu_j(x) < 0\},$$

$$C := \{x \in M_a^b \mid \text{there exists } m_\beta \in \beta(x) \text{ with } \sigma_{1,m_\beta}(x) \cdot \sigma_{2,m_\beta}(x) < 0\}.$$

Since each $\bar{x} \in M_a^b$ is not C-stationary for the MPCC, we get $\bar{x} \in A \cup B \cup C$.

The proof consists of a local argument and its globalization. First, we show the **local argument**. For each $\bar{x} \in M_a^b$, there exist an $(\mathbb{R}^n)$-neighborhood $U_{\bar{x}}$ of $\bar{x}$, $t_{\bar{x}} > 0$, and a mapping

$$\Psi^{\bar{x}} : \begin{cases} [0, t_{\bar{x}}] \times (M^b \cap U_{\bar{x}}) & \longrightarrow & M \\ (t, x) & \longmapsto & \Psi^{\bar{x}}(t, x) \text{ such that} \end{cases}$$

(i) $\Psi^{\bar{x}}(t, M^b \cap U_{\bar{x}}) \subset M^{b-t}$ for all $t \in [0, t_{\bar{r}})$,

(ii) $\Psi^{\bar{x}}(t_1 + t_2, \cdot) = \Psi^{\bar{x}}(t_1, \Psi^{\bar{x}}(t_2, \cdot))$ for all $t_1, t_2 \in [0, t_{\bar{x}})$ with $t_1 + t_2 \in [0, t_{\bar{x}})$,

(iii)   if $\bar{x} \in A \cup B$, then $\Psi^{\bar{x}}(\cdot, \cdot)$ is a C^1-flow corresponding to a C^1-vector field $F^{\bar{x}}$, and

(iv)   if $\bar{x} \in C$, then $\Psi^{\bar{x}}(\cdot, \cdot)$ is a Lipschitz flow.

Obviously, the level sets of f are mapped locally onto the level sets of $f \circ \Phi^{-1}$, where Φ is a C^1-diffeomorphism according to Definition 13. Applying the standard diffeomorphism Φ from Definition 14, we consider $f \circ \Phi^{-1}$ (denoted by f again). Thus, we have $\bar{x} = 0$ and f is given on the feasible set $\{0_s\} \times \mathbb{H}^{|J_0(\bar{x})|} \times \left(\partial \mathbb{H}^2\right)^{|\beta(\bar{x})|} \times \mathbb{R}^p$.

Case (a): $\bar{x} \in A$

Then, from Remark 10 there exists $l \in \{1, \ldots, p\}$ with $\frac{\partial f}{\partial x_l}(\bar{x}) \neq 0$. Define a local C^1-vector field $F^{\bar{x}}$ as

$$F^{\bar{x}}(x_1, \ldots, x_l, \ldots, x_n) := \left(0, \ldots, -\frac{\partial f}{\partial x_l}(x) \cdot \left(\frac{\partial f}{\partial x_l}(x)\right)^{-2}, \ldots, 0\right)^T.$$

After respective inverse changes of local coordinates, $F^{\bar{x}}$ induces the flow $\Psi^{\bar{x}}$, which fits the local argument (see Theorem 2.7.6 of [63] for details).

Case (b): $\bar{x} \in B$

Then, from Remark 10, there exists $j \in J_0(x)$ with $\frac{\partial f}{\partial x_j}(\bar{x}) < 0$. By means of a C^1-coordinate transformation (along the lines of Theorem 3.2.26 of [63]) in the j-th coordinate on $\mathbb{H}$, leaving the other coordinates unchanged, we obtain locally for f

$$f(x_1, \ldots, x_j, \ldots, x_n) = -x_j + f(x_1, \ldots, \bar{x}_j, \ldots, x_n).$$

Define a local C^1-vector field $F^{\bar{x}}$ as

$$F^{\bar{x}}(x_1, \ldots, x_j, \ldots, x_n) := (0, \ldots, 1, \ldots, 0)^T.$$

After respective inverse changes of local coordinates, $F^{\bar{x}}$ induces the flow $\Psi^{\bar{x}}$, which fits the local argument (see Theorem 3.3.25 of [63] for details).

Case (c): $\bar{x} \in C$

Then, from Remark 10 there exists $m_\beta \in \beta(x)$ with

$$\frac{\partial f}{\partial x_{1, m_\beta}}(\bar{x}) \cdot \frac{\partial f}{\partial x_{2, m_\beta}}(\bar{x}) < 0.$$

Without loss of generality, we assume that $\frac{\partial f}{\partial x_{1, m_\beta}}(\bar{x}) < 0$ and $\frac{\partial f}{\partial x_{2, m_\beta}}(\bar{x}) > 0$.

From the proof of Theorem 19, formula (2.20), we can obtain for f in new C^1-coordinates the representation

$$f(x_1, \ldots, x_j, \ldots, x_n) = -x_{1,m_\beta} + x_{2,m_\beta} + f(x_1, \ldots, \bar{x}_{1,m_\beta}, \bar{x}_{2,m_\beta}, \ldots, x_n).$$

Define the mapping $\Psi^{\bar{x}}$ locally as

$$\Psi^{\bar{x}}(t, x_1, \ldots, x_{1,m_\beta}, x_{2,m_\beta}, \ldots, x_n) :=$$

$$(x_1, \ldots, x_{1,m_\beta} + \max\{0, t - x_{2,m_\beta}\}, \max\{0, x_{2,m_\beta} - t\}, \ldots, x_n)^T.$$

After respective inverse changes of local coordinates, $\Psi^{\bar{x}}$ fits the local argument.

Note that in all of Cases (a)–(c), $\Psi^{\bar{x}}(t, \cdot)$ leaves the feasible set $\{0_s\} \times \mathbb{H}^{|J_0(\bar{x})|} \times (\partial \mathbb{H}^2)^{|\beta(\bar{x})|} \times \mathbb{R}^p$ invariant.

Globalization. Consider the open covering $\{U_x \mid x \in C\} \cup \{U_{\bar{x}} \mid \bar{x} \in M_a^b \backslash \{U_x \mid x \in C\}\}$ of M_a^b. From continuity arguments, $U_{\bar{x}}$, $\bar{x} \in M_a^b \backslash \{U_x \mid x \in C\}$ can be taken smaller, if necessary, to be disjoint with C. Since M_a^b is compact, we get a finite open subcovering $\{U_{x_i} \mid x_i \in C\} \cup \{U_{\bar{x}_j} \mid \bar{x}_j \in M_a^b \backslash \{U_x \mid x \in C\}\}$ of M_a^b. Using a C^∞-partition of unity $\{\phi_j\}$ subordinate to $\{U_{\bar{x}_j} \mid \bar{x}_j \in M_a^b \backslash \{U_x \mid x \in C\}\}$, we define with $F^{\bar{x}_j}$ (see Cases (a) and (b)) a C^1-vector field $F := \sum_j \phi_j F^{\bar{x}_j}$. The last induces a flow Ψ on $\{U_{\bar{x}_j} \mid \bar{x}_j \in M_a^b \backslash \{U_x \mid x \in C\}\}$ (see Theorem 3.3.14 of [63] for details). Note that in each nonempty overlapping region $U_{x_i} \cap U_{x_j}$, $x_i \in C$, $x_j \in M_a^b \backslash \{U_x \mid x \in C\}$, the flow Ψ^{x_i} induces exactly the vector field F (see Case (c)). Hence, local trajectories can be glued together on M_a^b, named by Ψ again. Moreover, moving along the local pieces of the trajectories $\Psi(\cdot, x)$, $x \in M_a^b$ reduces the level of f at least by a positive real number

$$\frac{\min\{t_{x_i}, t_{x_j} \mid x_i \in C, x_j \in M_a^b \backslash \{U_x \mid x \in C\}\}}{2}.$$

Thus, we obtain for $x \in M_a^b$ a unique $t_a(x) > 0$ with $\Psi(t_a(x), x) \in M^a$. It is not hard (but technical) to realize that $t_a : x \longrightarrow t_a(x)$ is Lipschitz. Finally, we define $r : [0,1] \times M^b \longrightarrow M^b$ as

$$r(\tau, x) : \begin{cases} x & \text{for } x \in M^a, \ \tau \in [0,1] \\ \Psi(\tau t_a(x), x) & \text{for } x \in M_a^b, \ \tau \in [0,1]. \end{cases}$$

The mapping r provides that M^a is a strong deformation retract of M^b.

(b) By virtue of the deformation theorem and the normal forms (2.19), (2.23) and (2.26), the proof of the cell-attachment part becomes standard. In fact, the deformation theorem allows deformations up to an arbitrarily small neighborhood of the C-stationary point $\bar{x}$. In such a neighborhood, we can work in continuous local coordinates and use the explicit normal form (2.26). In the normal form (2.26), the origin is a nondegenerate KKT point and the cell attachment can be performed as in Theorem 3.3.33 of [63]. $\square$

Remark 11. We emphasize that the linear terms y_i, $i \in J_0(\bar{x})$, in (2.26) do not contribute to the dimension of the cell to be attached. In fact, w.r.t. lower-level sets, the one-dimensional constrained singularity y, $y \geq 0$, plays the same role as the unconstrained singularity y^2. In this sense, the constrained linear terms in (2.26) do not contribute to the number of negative squares.

Remark 12. Another way of looking at the cell-attachment part is via stratified Morse theory (Section 3.7 of [29]). In fact, recall the normal form (2.19). The set $\{0_s\} \times \mathbb{H}^{|J_0(\bar{x})|} \times \left(\partial \mathbb{H}^2\right)^{|\beta(\bar{x})|} \times \mathbb{R}^p$ can be interpreted as the product of the "tangential part" $\{0_s\} \times \mathbb{R}^p$ and the "normal part" $\mathbb{H}^{|J_0(\bar{x})|} \times \left(\partial \mathbb{H}^2\right)^{|\beta(\bar{x})|}$. The main theorem in [29] states that the local "Morse data" is the product of the tangential "Morse data" with the normal "Morse data". The tangential Morse index equals QI and, in view of Remark 11, the normal Morse index equals BI. In the product, the index then becomes the sum QI+BI, which is precisely the C-index (see Figure 15).

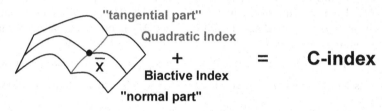

Figure 15 C-index

Remark 13. As pointed out by an anonymous referee, Theorem 20 can be interpreted as follows. The complementarity constraints can be reformulated as Lipschitzian equality constraints of the minimum type. For $u, v \in \mathbb{R}$, we have

$$u \geq 0, v \geq 0, u \cdot v = 0 \Longleftrightarrow \min\{u, v\} = 0.$$

Regarding this issue, Corollary 3 provides a normal form of f in Lipschitzian coordinates. Finally, Theorem 20 shows why the Morse index from the smooth nonlinear programming has to be modified into the Clarke index for the MPCC.

Discussion of different stationarity concepts

We briefly review well-known definitions of various stationarity concepts and connections between them (see [24], [96], [105]).

Definition 15. Let $\bar{x} \in M$.

(i) $\bar{x}$ is called W-stationary if (2.13) and (2.14) hold.

(ii) $\bar{x}$ is called A-stationary if (2.13) and (2.14) hold and

$$\bar{\sigma}_{1,m_\beta} \geq 0 \text{ or } \bar{\sigma}_{2,m_\beta} \geq 0 \text{ for all } m_\beta \in \beta(\bar{x}).$$

(iii) $\bar{x}$ is called M-stationary if (2.13) and (2.14) hold and

$$(\bar{\sigma}_{1,m_\beta} > 0 \text{ and } \bar{\sigma}_{1,m_\beta} > 0) \text{ or } \bar{\sigma}_{1,m_\beta} \cdot \bar{\sigma}_{2,m_\beta} = 0 \text{ for all } m_\beta \in \beta(\bar{x}).$$

(iv) $\bar{x}$ is called S-stationary if (2.13) and (2.14) hold and

$$\bar{\sigma}_{1,m_\beta} \geq 0, \bar{\sigma}_{2,m_\beta} \geq 0 \text{ for all } m_\beta \in \beta(\bar{x}).$$

(v) $\bar{x}$ is called B-stationary if $d = 0$ is a local solution of the linearized problem

$$\min f(\bar{x}) + Df(\bar{x})d \text{ s.t.}$$

$$\begin{cases} F_{1,m}(\bar{x}) + DF_{1,m}(\bar{x})d \geq 0, F_{2,m}(\bar{x}) + DF_{2,m}(\bar{x})d \geq 0, \\ (F_{1,m}(\bar{x}) + DF_{1,m}(\bar{x})d) \cdot (F_{2,m}(\bar{x}) + DF_{2,m}(\bar{x})d) = 0, m = 1, \ldots, k, \\ h(\bar{x}) + Dh(\bar{x})d = 0, g(\bar{x}) + Dg(\bar{x})d \geq 0. \end{cases}$$

The following diagram (see Figure 16) summarizes the relations between the stationarity concepts mentioned (e.g., [118]):

S-stationary point $\Longleftrightarrow$ **B-stationary point**

$\Downarrow$ under LICQ

M-stationary point

$\Downarrow$ $\Downarrow$

C-stationary point   **A-stationary point**

$\Downarrow$ $\Downarrow$

W-stationary point

Figure 16 Stationarity concepts in MPCC

Assuming nondegeneracy (as in Definition 11), we see that A-, M-, S-, and B-stationary points describe local minima tighter than C-stationary points. However, they exclude C-stationary points with $BI > 0$. These points are also crucial for the topological structure of the MPCC (see the cell-attachment theorem). For global optimization, points of *C-index* $= 1$ play an important role; see also Section 1.2. We emphasize that among the points of *C-index* $= 1$ from a topological point of view there is no substantial difference between the points with $BI = 1, QI = 0$ and $BI = 0, QI = 1$. It is worth mentioning that a linear descent direction might exist in a nondegenerate C-stationary point with positive *C-index* (see [87] and [105] and the following discussion). However, at points with $BI = 1, QI = 0$, there are exactly two directions of linear decrease. Both of them are important from a global point of view. In turn, W-stationary points contain those with negative and positive Lagrange multipliers corresponding to the same complementarity constraint. Rrom the deformation theorem, such points are irrelevant for the topological structure of the MPCC.

Furthermore, we illustrate the foregoing considerations by Example 12 from [105] (see also [87]).

Example 12.

$$\min (x - 1)^2 + (y - 1)^2 \text{ s.t. } x \geq 0, y \geq 0, x \cdot y = 0. \qquad (2.27)$$

It is clear that C-stationary points for (2.27) are $(1,0)$, $(1,0)$, and $(0,0)$. Moreover, $(1,0)$ and $(1,0)$ are local (and global) minimizers with *C-index* 0. The biactive Lagrange multipliers for $(0,0)$ are both -2; hence, its *C-index* is 1. One might think that the C-stationary point $(0,0)$ is irrelevant for numerical purposes since it possesses linear descent directions. However, globally it precisely connects the local minima. Moreover, if we consider the problem (2.27) with smoothed complementarity constraints,

$$\min (x-1)^2 + (y-1)^2 \text{ s.t. } x \geq 0, y \geq 0, x \cdot y = \varepsilon, \qquad (2.28)$$

where $\varepsilon > 0$ is sufficiently small. Then, it is easily seen that the critical points for (2.28) are

$$(x_1, y_1) = \left(\frac{1+\sqrt{1-4\varepsilon}}{2}, \frac{1-\sqrt{1-4\varepsilon}}{2} \right),$$

$$(x_2, y_2) = \left(\frac{1-\sqrt{1-4\varepsilon}}{2}, \frac{1+\sqrt{1-4\varepsilon}}{2} \right),$$

$$(x_3, y_3) = \left(\sqrt{\varepsilon}, \sqrt{\varepsilon} \right).$$

Obviously, $(x_1, y_1) \longrightarrow (1,0)$, $(x_2, y_2) \longrightarrow (0,1)$, and $(x_3, y_3) \longrightarrow (0,0)$ as $\varepsilon \longrightarrow 0$. Moreover, (x_1, y_1) and (x_2, y_2) are local (and global) minimizers for (2.28) with quadratic index 0, and the quadratic index of (x_3, y_3) is 1 (local maximum). Hence, by the smoothing procedure, the C-stationary point $(0,0)$ with *C-index* 1 corresponds to the critical point (x_3, y_3) with quadratic index 1. In particular, the smoothed version preserves the global topological structure.

We notice that adding positive squares to the objective function in (2.27) provides a higher-dimensional example with the same features.

2.4 Parametric aspects

The aim of this section is the introduction and characterization of the strong stability notion in the MPCC (see Definition 17). In 1980, M. Kojima introduced in [83] the (topological) concept of strong stability of stationary solutions (Karush-Kuhn-Tucker points) for nonlinear programming (see also Robinson [101] and Section 1.4). This concept plays an important role in optimization theory, for example in sensitivity and parametric optimization [64, 84] and structural stability [78]. It turns out that the concept of C-stationarity is an adequate stationarity concept regarding possible bifurcations.

We characterize the strong stability for C-stationary points by means of first- and second-order information of the defining functions f, h, g, F_1, F_2 under the LICQ (see Theorem 21). The main issue in the strong stability of C-stationary points is related to the so-called biactive Lagrange multipliers (see also Corollary 5). A biactive pair of Lagrange multipliers corresponds to such complementarity constraints which both vanish at a C-stationary point. There are three (degeneracy) possibilities for biactive multipliers:

(a) Both biactive Lagrange multipliers do not vanish (nondegenerate case).
(b) Only one biactive Lagrange multiplier vanishes (first degenerate case).
(c) Both biactive Lagrange multipliers vanish (second degenerate case).

Depending on the kind of possible degeneracy, we use corresponding ideas on strong stability of Kojima (Cases (a) and (b)). Moreover, we describe new unstable phenomena (Case (c)).

We would like to refer to some related papers. In [105], an extension of the stability results of Kojima and Robinson to the MPCC is presented. It refers to the nondegenerate Case (a) of nonvanishing biactive Lagrange multipliers. In [99], the concept of the so-called co-1-singularity for quadratic programs with complementarity constraints (QPCCs) is studied. In our terms they refer to the first degenerate Case (b). We refer the reader to [76] for details.

Notation and Auxiliary Results

From Section 2.3 we recall the following index sets given $\bar{x} \in M$:

$$J_0(\bar{x}) := \{ j \in J \mid g_j(\bar{x}) = 0 \},$$
$$\alpha(\bar{x}) := \{ m \in \{1, \ldots k\} \mid F_{1,m}(\bar{x}) = 0, F_{2,m}(\bar{x}) > 0 \},$$
$$\beta(\bar{x}) := \{ m \in \{1, \ldots k\} \mid F_{1,m}(\bar{x}) = 0, F_{2,m}(\bar{x}) = 0 \},$$
$$\gamma(\bar{x}) := \{ m \in \{1, \ldots k\} \mid F_{1,m}(\bar{x}) > 0, F_{2,m}(\bar{x}) = 0 \}.$$

Without loss of generality, we assume that at the particular point of interest $\bar{x} \in M$ it holds that

$$J_0(\bar{x}) = \{1, \ldots, |J_0(\bar{x})|\}, \; \alpha(\bar{x}) = \{1, \ldots, |\alpha(\bar{x})|\},$$
$$\gamma(\bar{x}) = \{|\alpha(\bar{x})| + 1, \ldots, |\alpha(\bar{x})| + |\gamma(\bar{x})|\}.$$

We put $s := |I| + |\alpha(\bar{x})| + |\gamma(\bar{x})|$, $p := n - s - |J_0(\bar{x})| - 2|\beta(\bar{x})|$.

We also recall the notions of the LICQ and C-stationarity (see Section 2.3).

The LICQ for MPCC is said to hold at $\bar{x} \in M$ if the vectors

$$Dh_i(\bar{x}), \; i \in I, DF_{1,m_\alpha}(\bar{x}), \; m_\alpha \in \alpha(\bar{x}), DF_{2,m_\gamma}(\bar{x}), \; m_\gamma \in \gamma(\bar{x}),$$
$$Dg_j(\bar{x}), \; j \in J_0(\bar{x}), DF_{1,m_\beta}(\bar{x}), DF_{2,m_\beta}(\bar{x}), \; m_\beta \in \beta(\bar{x})$$

are linearly independent.

A point $\bar{x} \in M$ is called C-stationary for the MPCC (see Definition 10) if there exist real numbers $\bar{\lambda}_i, \; i \in I, \bar{\mu}_j, \; j \in J, \bar{\sigma}_{1,m}, \bar{\sigma}_{2,m}, \; m = 1, \ldots, k$ (Lagrange multipliers) such that

$$Df(\bar{x}) = \sum_{i \in I} \bar{\lambda}_i Dh_i(\bar{x}) + \sum_{j \in J} \bar{\mu}_j Dg_j(\bar{x})$$

$$+ \sum_{m=1}^{k} \left(\bar{\sigma}_{1,m} D\Gamma_{1,m}(\bar{x}) \mid \bar{\sigma}_{2,m} DF_{2,m}(\bar{x}) \right), \tag{2.29}$$

$$\bar{\mu}_j \cdot g_j(\bar{x}) = 0, \; j \in J, \tag{2.30}$$

$$\bar{\mu}_j \geq 0 \text{ for all } j \in J_0(\bar{x}), \tag{2.31}$$

$$\bar{\sigma}_{j,m} \cdot F_{j,m}(\bar{x}) = 0, \; j = 1, 2, \; m = 1, \ldots, k, \tag{2.32}$$

$$\bar{\sigma}_{1,m_\beta} \cdot \bar{\sigma}_{2,m_\beta} \geq 0 \text{ for all } m_\beta \in \beta(\bar{x}). \tag{2.33}$$

The Lagrange function L is defined as follows:

$$L(x, \lambda, \mu, \sigma) := f(x) - \sum_{i \in I} \lambda_i h_i(\bar{x}) - \sum_{j \in J} \mu_j g_j(\bar{x})$$

$$- \sum_{m=1}^{k} \left(\sigma_{1,m} F_{1,m}(\bar{x}) + \sigma_{2,m} F_{2,m}(\bar{x}) \right). \tag{2.34}$$

Definition 16 (C-stationary pair). A vector $(\bar{x}, \bar{\lambda}, \bar{\mu}, \bar{\sigma}) \in M \times \mathbb{R}^{|I|} \times \mathbb{R}^{|J|} \times \mathbb{R}^{2k}$ satisfying (2.29)–(2.33) is called a C-stationary pair for the MPCC.

The concept of strong stability is defined by means of an appropriate seminorm. To this aim, let $\bar{x} \in \mathbb{R}^n$, $r > 0$. For defining functions (f, h, g, F_1, F_2) from (2.5), the seminorm $\|(f, h, g, F_1, F_2)\|_{B(\bar{x},r)}^{C^2}$ is defined to be the maximum modulus of the function values and partial derivatives up to order 2 of f, h, g, F_1, F_2.

Definition 17 (Strong stability). A C-stationary point $\bar{x} \in M$ (resp. a C-stationary pair $(\bar{x}, \bar{\lambda}, \bar{\mu}, \bar{\sigma})$), for $MPCC[f, g, h, F_1, F_2]$ is called (C^2)-strongly stable if for some $r > 0$ and each $\varepsilon \in (0, r]$ there exists $\delta = \delta(\varepsilon) > 0$ such that whenever

$$\left(\widetilde{f}, \widetilde{h}, \widetilde{g}, \widetilde{F}_1, \widetilde{F}_2 \right) \in C^2$$

and

$$\left\| \left(f - \widetilde{f}, h - \widetilde{h}, g - \widetilde{g}, F_1 - \widetilde{F}_1, F_2 - \widetilde{F}_2 \right) \right\|_{B(\bar{x},r)}^{C^2} \leq \delta,$$

$B(\bar{x}, \varepsilon)$ (resp. $B\left((\bar{x}, \bar{\lambda}, \bar{\mu}, \bar{\sigma}), \varepsilon \right)$) contains a C-stationary point

$$\widetilde{x} \left(\text{resp. a C-stationary pair } \left(\widetilde{x}, \widetilde{\lambda}, \widetilde{\mu}, \widetilde{\sigma} \right) \right) \text{ for MPCC} \left[\widetilde{f}, \widetilde{h}, \widetilde{g}, \widetilde{F}_1, \widetilde{F}_2 \right]$$

that is unique in $B(\bar{x}, r)$ $\left(\text{resp. unique in } B\left(\left(\widetilde{x}, \widetilde{\lambda}, \widetilde{\mu}, \widetilde{\sigma} \right), r \right) \right)$.

The following lemma establishes the connection between both definitions just introduced (see [82]).

Lemma 6 (C-stationary points and pairs). *The following assertions are equivalent:*

(a) $\bar{x}$ *is a strongly stable C-stationary point for the MPCC that satisfies the LICQ,*
   and $(\bar{\lambda},\bar{\mu},\bar{\sigma})$ *is the associated Lagrange multiplier vector.*
(b) $(\bar{x},\bar{\lambda},\bar{\mu},\bar{\sigma})$ *is a strongly stable C-stationary pair for the MPCC.*

Proof. (a) $\Longrightarrow$ (b) The LICQ remains valid under small perturbations of the defining functions. Hence, the corresponding Lagrange multipliers are unique. Moreover, Remark 10 provides the continuity of Lagrange multipliers w.r.t. perturbations under consideration.

(b) $\Longrightarrow$ (a) The nontrivial part is to prove that LICQ holds at $\bar{x}$. The proof goes along the lines of Theorem 2.3 from [82]. To stress the new aspects, here we assume that there are only biactive constraints (i.e., $I = \emptyset$, $J = \emptyset$, $\alpha(\bar{x}) = \emptyset$, and $\gamma(\bar{x}) = \emptyset$). Let $(\bar{x},\bar{\sigma})$ be a strongly stable C-stationary pair for the MPCC and let the LICQ not be fulfilled at $\bar{x}$. Then, there exist real numbers δ_{1,m_β}, δ_{1,m_β}, $m_\beta \in \beta(\bar{x})$ (not all vanishing) such that

$$\sum_{m_\beta \in \beta(\bar{x})} (\delta_{1,m}DF_{1,m}(\bar{x}) + \delta_{2,m}DF_{2,m}(\bar{x})) = 0. \tag{2.35}$$

Setting

$$m_\beta^+(\bar{x}) := \{m_\beta \in \beta(\bar{x}) \mid \bar{\sigma}_{1,m}, \bar{\sigma}_{2,m} \geq 0\},$$

$$m_\beta^-(\bar{x}) := \{m_\beta \in \beta(\bar{x}) \mid \bar{\sigma}_{1,m}, \bar{\sigma}_{2,m} \leq 0\},$$

we define

$$c := -\left[\sum_{m_\beta \in m_\beta^+(\bar{x})} (DF_{1,m}(\bar{x}) + DF_{2,m}(\bar{x})) - \sum_{m_\beta \in m_\beta^-(\bar{x})} (DF_{1,m}(\bar{x}) + DF_{2,m}(\bar{x}))\right].$$

For $\varepsilon > 0$, let

$$\sigma_{1,m}(\varepsilon) := \bar{\sigma}_{1,m} + \varepsilon, \ \sigma_{2,m}(\varepsilon) := \bar{\sigma}_{2,m} + \varepsilon \text{ for all } m \in m_\beta^+(\bar{x}),$$

$$\sigma_{1,m}(\varepsilon) := \bar{\sigma}_{1,m} - \varepsilon, \ \sigma_{2,m}(\varepsilon) := \bar{\sigma}_{2,m} - \varepsilon \text{ for all } m \in m_\beta^-(\bar{x}).$$

Putting $\varphi(x) := c \cdot x$, we obtain that

$$(\bar{x},\sigma(\varepsilon)) \text{ is a C-stationary pair for MPCC}[f + \varepsilon \cdot \varphi, F_1, F_2].$$

Moreover, due to the strong stability of $(\bar{x},\bar{\sigma})$ for MPCC$[f,F_1,F_2]$, we claim that for each sufficiently small $\varepsilon > 0$ the C-stationary pair $(\bar{x},\sigma(\varepsilon))$ is unique for MPCC$[f + \varepsilon \cdot \varphi, F_1, F_2]$ in some neighborhood U of $(\bar{x},\bar{\sigma})$.

However, (2.35) and $\sigma_{i,m}(\varepsilon) \neq 0$ for $m \in m_\beta(\bar{x})$, $i = 1,2$ ensure that, for any sufficiently small real number t, the pair $(\bar{x},v(\varepsilon,\delta,t))$ with

$$v_{1,m}(\varepsilon,\delta,t) := \sigma_{1,m}(\varepsilon) + \delta_{1,m}t, \ v_{2,m}(\varepsilon,\delta,t) := \sigma_{2,m}(\varepsilon) + \delta_{2,m}t \text{ for all } m \in m_\beta(\bar{x})$$

belongs to U and is a C-stationary pair for $\text{MPCC}[f + \varepsilon \cdot \varphi, F_1, F_2]$. Hence, necessarily $\delta = 0$, and the LICQ is shown. $\square$

Now we give two guiding examples for instability that may occur at C-stationary points in the second degenerate Case (c).

Example 13 (Unstable minimum/maximum [105]). Consider the MPCC:

$$\min x^2 + y^2 \text{ s.t. } x \geq 0, y \geq 0, x \cdot y = 0. \tag{2.36}$$

Obviously, $(0,0)$ is the unique C-stationary point for (2.36) with both vanishing biactive Lagrange multipliers. Consider the following perturbation of (2.36) w.r.t. parameter $t > 0$:

$$\min (x - t)^2 + (y - t)^2 \text{ s.t. } x \geq 0, y \geq 0, x \cdot y = 0. \tag{2.37}$$

It is easy to see that $(0,0)$, $(0,t)$, and $(t,0)$ are C-stationary points for (2.37). This means that $(0,0)$ is not a strongly stable C-stationary point for (2.36). Analogously, we can treat $-x^2 - y^2$ on $\partial \mathbb{H}^2$ at the origin.

Example 14 (Unstable saddle point). Consider the MPCC

$$\min x^2 - y^2 \text{ s.t. } x \geq 0, y \geq 0, x \cdot y = 0. \tag{2.38}$$

Obviously, $(0,0)$ is the unique C-stationary point for (2.38) with both vanishing biactive Lagrange multipliers. Consider the following perturbation of (2.38) w.r.t. parameter $t > 0$:

$$\min (x - t)^2 - (y - t)^2 \text{ s.t. } x \geq 0, y \geq 0, x \cdot y = 0. \tag{2.39}$$

It is easy to see that $(0,t)$ and $(t,0)$ are C-stationary points for (2.39). This means that $(0,0)$ is not a strongly stable C-stationary point for (2.38).

Characterization of strong stability for C-stationary points

Before stating the main result, we define the following index sets at a C-stationary point $\bar{x} \in M$ with Lagrange multipliers $(\bar{\lambda}, \bar{\mu}, \bar{\sigma})$ (see Definition 10):

$$J_+ := \{ j \in J_0(\bar{x}) \,|\, \bar{\mu}_j > 0 \},$$
$$p(\bar{x}) := \{ m \in \beta(\bar{x}) \,|\, \bar{\sigma}_{1,m} \cdot \bar{\sigma}_{2,m} > 0 \},$$
$$q(\bar{x}) := \{ m \in \beta(\bar{x}) \,|\, \bar{\sigma}_{1,m} > 0, \bar{\sigma}_{2,m} = 0 \},$$
$$r(\bar{x}) := \{ m \in \beta(\bar{x}) \,|\, \bar{\sigma}_{1,m} = 0, \bar{\sigma}_{2,m} > 0 \},$$
$$s(\bar{x}) := \{ m \in \beta(\bar{x}) \,|\, \bar{\sigma}_{1,m} < 0, \bar{\sigma}_{2,m} = 0 \},$$
$$w(\bar{x}) := \{ m \in \beta(\bar{x}) \,|\, \bar{\sigma}_{1,m} = 0, \bar{\sigma}_{2,m} < 0 \},$$

$$u(\bar{x}) := \{m \in \beta(\bar{x}) \mid \bar{\sigma}_{1,m} = 0, \bar{\sigma}_{2,m} = 0\}.$$

Obviously, $p(\bar{x})$, $q(\bar{x})$, $r(\bar{x})$, $s(\bar{x})$, $w(\bar{x})$, and $u(\bar{x})$ constitute a partition of $\beta(\bar{x})$.

For $\bar{J} \subset J$, $K \subset \{1,\ldots,k\}$, $j = 1,2$, we write h, $g_{\bar{J}}$, and $F_{j,K}$ for $(h_i \mid i \in I)$, $(g_j \mid j \in \bar{J})$, and $(F_{j,m} \mid m \in K)$, respectively.

Furthermore, for $J_+ \subset \bar{J} \subset J_0(\bar{x})$, $\bar{q} \subset q(\bar{x})$, $\bar{r} \subset r(\bar{x})$, $\bar{s} \subset s(\bar{x})$, $\bar{w} \subset w(\bar{x})$, we define $M_{\bar{J},\bar{q},\bar{r},\bar{s},\bar{w}}(\bar{x})$ to be the block matrix $\begin{pmatrix} C & X \\ Y & 0 \end{pmatrix}$ with

$$X = \begin{pmatrix} H^T & G_{\bar{J}}^T & A^T & \Gamma^T & P^T & Q^T & \bar{Q}^T & R^T & \bar{R}^T & S^T & \bar{S}^T & W^T & \bar{W}^T \end{pmatrix},$$

$$Y^T = \begin{pmatrix} H^T & -G_{\bar{J}}^T & A^T & \Gamma^T & P^T & Q^T & -\bar{Q}^T & R^T & -\bar{R}^T & S^T & \bar{S}^T & W^T & \bar{W}^T \end{pmatrix},$$

where

$$C = D_{xx}^2 L(\bar{x}, \bar{\lambda}, \bar{\mu}, \bar{\sigma}), H = Dh(\bar{x}), G_{\bar{J}} = Dg_{\bar{J}}(\bar{x}),$$

$$A = DF_{1,\alpha(\bar{x})}(\bar{x}), \Gamma = DF_{2,\gamma(\bar{x})}(\bar{x}), P = \left(DF_{1,p(\bar{x})}, DF_{2,p(\bar{x})} \right)(\bar{x}),$$

$$Q = DF_{1,q(\bar{x})}(\bar{x}), \bar{Q} = DF_{2,\bar{q}}(\bar{x}), R = DF_{2,r(\bar{x})}(\bar{x}), \bar{R} = DF_{1,\bar{r}}(\bar{x}),$$

$$S = DF_{1,s(\bar{x})}(\bar{x}), \bar{S} = DF_{2,\bar{s}}(\bar{x}), W = DF_{2,w(\bar{x})}(\bar{x}), \bar{W} = DF_{1,\bar{w}}(\bar{x}).$$

Theorem 21 (Characterization of strong stability). *Suppose that the LICQ holds at a C-stationary point $\bar{x} \in M$ with Lagrange multipliers $(\bar{\lambda}, \bar{\mu}, \bar{\sigma})$ (see Definition 10). Then, $\bar{x}$ is a strongly stable C-stationary point for the MPCC (see Definition 17) if and only if*

(i) $u(\bar{x}) = \emptyset$ *and*
(ii)   *all matrices $M_{\bar{J},\bar{q},\bar{r},\bar{s},\bar{t}}(\bar{x})$ with*

$$J_+ \subset \bar{J} \subset J_0(\bar{x}), \bar{q} \subset q(\bar{x}), \bar{r} \subset r(\bar{x}), \bar{s} \subset s(\bar{x}), \bar{w} \subset w(\bar{x})$$

are nonsingular with the same determinant sign.

Proof. By virtue of the LICQ at $\bar{x}$, Lemma 6 allows us to deal equivalently with the strong stability of the C-stationary pair $(\bar{x}, \bar{\lambda}, \bar{\mu}, \bar{\sigma})$.

Case 1: $u(\bar{x}) = \emptyset$

We consider the following mapping $\mathscr{T} : \mathbb{R}^n \times \mathbb{R}^{|I|} \times \mathbb{R}^{|J|} \times \mathbb{R}^{2k} \longrightarrow \mathbb{R}^{n+|I|+|J|+2k}$ locally at its zero $(\bar{x}, \bar{\lambda}, \bar{\mu}, \bar{\sigma})$:

$$\mathscr{T}(x,\lambda,\mu,\sigma) := \begin{pmatrix} D_x L(x,\lambda,\mu,\sigma) \\ h(x) \\ \min\{\mu, g(x)\} \\ F_{1,\alpha(\bar{x})}(x) \\ F_{2,\gamma(\bar{x})}(x) \\ F_{1,p(\bar{x})}(x) \\ F_{2,p(\bar{x})}(x) \\ F_{1,q(\bar{x})}(x) \\ \min\{\sigma_{2,q(\bar{x})}, F_{2,q(\bar{x})}(x)\} \\ F_{2,r(\bar{x})} \\ \min\{\sigma_{1,r(\bar{x})}, F_{1,r(\bar{x})}(x)\} \\ F_{1,s(\bar{x})} \\ \min\{-\sigma_{2,s(\bar{x})}, F_{2,s(\bar{x})}(x)\} \\ F_{2,w(\bar{x})} \\ \min\{-\sigma_{1,w(\bar{x})}, F_{1,w(\bar{x})}(x)\} \end{pmatrix}.$$

Note that C-stationary pairs for MPCC in a sufficiently small neighborhood of $(\bar{x}, \bar{\lambda}, \bar{\mu}, \bar{\sigma})$, are precisely the zeros of $\mathscr{T}$. Moreover, the only difference in $\mathscr{T}$ compared with the case of nonlinear programming is the appearing minus sign in

$$\min\{-\sigma_{2,s(\bar{x})}, F_{2,s(\bar{x})}(x)\}, \ \min\{-\sigma_{1,w(\bar{x})}, F_{1,w(\bar{x})}(x)\}.$$

Since we deal with equality constraints of minimum type, Theorem 4.3 from [82] (characterization of strong stability for KKT points) can be simply adapted here. Indeed, as in Theorem 4.3 from [82], the strong stability for $(\bar{x}, \bar{\lambda}, \bar{\mu}, \bar{\sigma})$ can be characterized by the fact that all matrices in the Clarke subdifferential $\partial \mathscr{T}(\bar{x}, \bar{\lambda}, \bar{\mu}, \bar{\sigma})$ are nonsingular. The latter can equivalently be rewritten as condition (ii) (see also [80] and Theorem 9 for the case of nonlinear programming).

Case 2: $\mathbf{u}(\bar{x}) \neq \emptyset$

Let $\Phi : U \longrightarrow V$ be a standard diffeomorphism according to Definition 14. We put $\bar{f} := f \circ \Phi^{-1}$ on the set $\left(\{0_s\} \times \mathbb{H}^{|J_0(\bar{x})|} \times (\partial \mathbb{H}^2)^{|\beta(\bar{x})|} \times \mathbb{R}^p\right) \cap V$. From now on, we may assume $s = 0$. In view of Remark 10, we have at the origin

(i) $\dfrac{\partial \bar{f}}{\partial y_j} \geq 0, \ j \in J_0(\bar{x}),$

(ii) $\dfrac{\partial \bar{f}}{\partial y_{|J_0|+2m-1}} \cdot \dfrac{\partial \bar{f}}{\partial y_{|J_0|+2m}} \geq 0, \ m = 1, \dots |\beta(\bar{x})|,$

(iii) $\dfrac{\partial \bar{f}}{\partial y_{l+n-p}} = 0, \ l = 1, \dots, p.$

Moreover, due to condition $u(\bar{x}) \neq \emptyset$ we may assume w.l.o.g. that

(iv) $\dfrac{\partial \bar{f}}{\partial y_{|J_0|+1}} = 0, \ \dfrac{\partial \bar{f}}{\partial y_{|J_0|+2}} = 0.$

From now on, we denote $\bar{f}$ again by f.

In what follows, we successively perform arbitrarily small perturbations of f such that the origin remains a C-stationary point on $\mathbb{H}^{|J_0(\bar{x})|} \times \left(\partial \mathbb{H}^2\right)^{|\beta(\bar{x})|} \times \mathbb{R}^p$.

(1) As a stabilization step, we add to f an arbitrarily small linear-quadratic term

$$\sum_{j=1}^{|J_0(\bar{x})|} c_j \cdot y_j + \sum_{m=2}^{|\beta(\bar{x})|} \left(c_{|J_0|+2m-1} \cdot y_{|J_0|+2m-1} + c_{|J_0|+2m} \cdot y_{|J_0|+2m}\right) + \sum_{l=1}^{p} c_{l+n-p} y_{l+n-p}^2$$

such that it holds that for the perturbed function (denoted again by f)

(i) $\quad \dfrac{\partial f}{\partial y_j} > 0,\ j \in J_0(\bar{x})$,

(ii) $\quad \dfrac{\partial f}{\partial y_{|J_0|+2m-1}} \cdot \dfrac{\partial f}{\partial y_{|J_0|+2m}} > 0,\ m = 2,\ldots |\beta(\bar{x})|$,

(iii) $\quad \dfrac{\partial f}{\partial y_{l+n-p}} = 0,\ l = 1,\ldots,p$, and

$$\left(\frac{\partial^2 f}{\partial y_{k_1+n-p}\partial y_{k_2+n-p}}\right)_{1 \leq k_1,k_2 \leq p} \quad \text{is a nonsingular matrix,}$$

and

(iv) $\quad \dfrac{\partial f}{\partial y_{|J_0|+1}} = 0,\ \dfrac{\partial f}{\partial y_{|J_0|+2}} = 0.$

(2) We approximate f by means of a C^∞-function in a small C^2-neighborhood of f leaving its value and its first- and second-order derivatives at the origin invariant. This can be done since C^∞-functions lie C^2-dense in C^2-functions. We denote the latter C^∞-approximation again by f.

From the stabilization step (1) and step (2), we can restrict our considerations to the following case:

$$f \in C^\infty\left(\mathbb{R}^2, \mathbb{R}\right),\ 0 \text{ is a C-stationary point for } f_{|\partial \mathbb{H}^2} \text{ and } \frac{\partial f}{\partial x}(0) = \frac{\partial f}{\partial y}(0) = 0.$$

Now we can write $f(x,y)$ as

$$f(x,y) = g_{1,1}(x,y)x^2 + 2g_{1,2}(x,y)xy + g_{2,2}(x,y)y^2$$

with $g_{1,1}, g_{1,2}, g_{2,2} \in C^\infty\left(\mathbb{R}^2, \mathbb{R}\right)$.

Adding to f an arbitrarily small quadratic term $ax^2 + by^2,\ a,b \in \mathbb{R}$, we can ensure that

$$g_{1,1}(0,0) \neq 0 \text{ and } g_{2,2}(0,0) \neq 0.$$

Hence, $\Psi(x,y) := \begin{pmatrix} x \cdot \sqrt{|g_{1,1}(x,y)|} \\ y \cdot \sqrt{|g_{2,2}(x,y)|} \end{pmatrix}$ is a local C^∞-diffeomorphism leaving $\partial \mathbb{H}^2$ invariant. In new local coordinates induced by Ψ, we obtain

$$f(x,y) = \varepsilon_1 x^2 + G(x,y)xy + \varepsilon_2 y^2,$$

where $\varepsilon_1 = \text{sign}(g_{1,1}(0,0))$, $\varepsilon_2 = \text{sign}(g_{2,2}(0,0))$.

Since $G(x,y)xy = 0$ on $\partial\mathbb{H}^2$, we can perturb f by means of a real parameter as in Example 13 or 14 to get a bifurcation of 0 as a C-stationary point.

Finally, performing all perturbations described above, we ensure that 0 is not a strongly stable C-stationary point. $\square$

The main new issue in the characterization of strong stability of C-stationary points in the MPCC will be clarified in Corollary 5. Its proof follows from Theorem 21 by means of a few elementary calculations.

Corollary 5. *Let $f \in C^2(\mathbb{R}^2, R)$ and suppose that 0 is a C-stationary point for $f_{|\partial\mathbb{H}^2}$. Then, 0 is a strongly stable C-stationary point if and only if*

$$\text{either } \frac{\partial f}{\partial x} \cdot \frac{\partial f}{\partial y} > 0 \text{ or } \frac{\partial f}{\partial x} = 0, \frac{\partial^2 f}{\partial x^2} \cdot \frac{\partial f}{\partial y} > 0 \text{ or } \frac{\partial f}{\partial y} = 0, \frac{\partial^2 f}{\partial y^2} \cdot \frac{\partial f}{\partial x} > 0 \text{ at } 0.$$

Now, we relate the notion of a nondegenerate C-stationary point to the results in Theorem 21 and Corollary 5.

Corollary 6. *Let $\bar{x} \in M$ be a nondegenerate C-stationary point as in Definition 11. Then, $\bar{x}$ is a strongly stable C-stationary point for the MPCC.*

In the situation of Corollary 5, we claim that 0 is a nondegenerate C-stationary point for $f_{|\partial\mathbb{H}^2}$ if and only if

$$\frac{\partial f}{\partial x} \cdot \frac{\partial f}{\partial y} > 0 \text{ at } 0.$$

From this, we get the following degeneracy possibilities of biactive Lagrange multipliers in the situation of Corollary 5 (see Figure 17):

(a)   nondegenerate case: both biactive Lagrange multipliers do not vanish,

$$\frac{\partial f}{\partial x} \cdot \frac{\partial f}{\partial y} > 0 \longrightarrow \textbf{stability},$$

(b)   first degenerate case: only one biactive Lagrange multiplier vanishes,

$$\frac{\partial f}{\partial x} = 0, \frac{\partial^2 f}{\partial x^2} \cdot \frac{\partial f}{\partial y} > 0 \text{ or } \frac{\partial f}{\partial y} = 0, \frac{\partial^2 f}{\partial y^2} \cdot \frac{\partial f}{\partial x} > 0 \longrightarrow \textbf{stability},$$

(c)   second degenerate case: both biactive Lagrange multipliers vanish,

$$\frac{\partial f}{\partial x} = 0, \frac{\partial f}{\partial y} = 0 \longrightarrow \textbf{instability}.$$

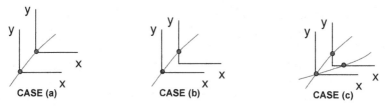

Figure 17 Strong stability in MPCC

On stability w.r.t. different stationarity concepts

For different stationarity concepts (such as A-, M-, S-, and B-stationarity), we refer the reader to Definition 15. Strong stability for A-, M-, S-, and B-stationary points can be defined analogously as in Definition 17. From now on, we assume that the LICQ holds at all points of interest.

It is clear that strongly stable S-stationary points can be characterized by means of Theorem 21. Indeed, each (not) strongly stable S-stationary point corresponds to a (not) strongly stable C-stationary point.

However, the issue becomes different as soon as we consider M-stationary points (or A-stationary points).

Example 15. Consider the MPCC

$$\min -x - y^2 \text{ s.t. } x \geq 0, y \geq 0, x \cdot y = 0. \tag{2.40}$$

Obviously, $(0,0)$ is the unique C-stationary point for (2.40) with biactive Lagrange multipliers $(-1,0)$. Hence, $(0,0)$ is also M-stationary. Moreover, from Corollary 5, $(0,0)$ is a strongly stable C-stationary point. Consider the following perturbation of (2.40) w.r.t. parameter $t > 0$:

$$\min -x - (y+t)^2 \text{ s.t. } x \geq 0, y \geq 0, x \cdot y = 0. \tag{2.41}$$

It is easy to see that $(0,0)$ is the unique C-stationary point for (2.41) with both biactive Lagrange multipliers $(-1, -2t)$ negative. It means that $(0,0)$ is not a strongly stable M-stationary point for (2.40).

Remark 14. We consider once more Example 13. We recall that $(0,0)$ is the unique C-stationary point for (2.36) with both vanishing biactive Lagrange multipliers. Hence, $(0,0)$ is also M-stationary. For the perturbed program (2.37), we have that $(0,0)$, $(0,t)$, and $(t,0)$ are C-stationary. It is easy to see that $(0,0)$ is not M-stationary for (2.37).

We note that adding positive or negative squares to the objective functions above provides higher-dimensional examples with similar features.

Finally, we point out some issues with the MPCC motivated by the strong stability results.

Remark 15. From Example 15 and Remark 14, the concept of C-stationarity is crucial for the numerical treatment of the MPCC via homotopy-based methods. Furthermore, Theorem 21 provides a characterization of the strongly stable C-stationary points. These are solutions of certain stable equations involving first- and second-order information of the defining functions. This fact might be used to establish some nonsmooth versions of the Newton method for the MPCC (see [53, 57]). This is an issue of current research.

Remark 16. In Theorem 21, the strong stability of C-stationary points is characterized under the LICQ. Its characterization in the absence of the LICQ is still open. We point out that this issue might be related to the version of the Mangasarian-Fromovitz condition (MFC) as introduced in [70] (see also Definition 7). The constraint qualification MFC has been introduced in connection with topological stability of the MPCC feasible set. This is still an issue of current research.

Chapter 3
General Semi-infinite Programming Problems

We study general semi-infinite programming problems (GSIP) from the topological point of view. Introducing the symmetric Mangasarian-Fromovitz constraint qualification (Sym-MFCQ) for GSIPs, we describe the closure of the GSIP feasible set. It is proved that Sym-MFCQ is stable and generic. Moreover, under Sym-MFCQ, the GSIP feasible set is shown to be a Lipschitz manifold. For GSIPs, we state the non-smooth symmetric reduction ansatz (NSRA). NSRA is proven to hold generically at all KKT points for the GSIP. NSRA allows us to reduce the GSIP to a so-called disjunctive optimization problem. This reduction enables to establish the critical point theory for GSIPs.

3.1 Applications and examples

Generalized semi-infinite programming problems (GSIPs) have the form

$$\text{GSIP:} \quad \text{minimize } f(x) \text{ s.t. } x \in M \tag{3.1}$$

with

$$M := \{x \in \mathbb{R}^n \mid g_0(x,y) \geq 0 \text{ for all } y \in Y(x)\}$$

and

$$Y(x) := \{y \in \mathbb{R}^m \mid g_k(x,y) \leq 0, k = 1, \ldots, s\}.$$

All defining functions $f, g_k, k = 0, \ldots, s$ are assumed to be real-valued and continuously differentiable on their respective domains. In the case of a constant mapping $Y(\cdot) = Y$, we refer to semi-infinite programming problems (SIPs).

We present the well-known applications of GSIPs in the area of Chebyshev approximation, design centering, and robust optimization from a survey [40]. Furthermore, we give some examples of GSIPs that illustrate two main new features of them (in addition to SIPs):

M need not be a closed set

and

M might exhibit so-called re-entrant corners.

Chebyshev and reverse Chebyshev approximations

We approximate a given continuous function F on a nonempty and compact set $Y \subset \mathbb{R}^m$ by a function $f(x, \cdot)$. The latter comes from a parameterized family of continuous functions $\{f(x, \cdot) \mid x \in X\}$ with some parameter set $X \subset \mathbb{R}^n$. The problem of Chebyshev approximation (CA) is as follows:

$$\text{CA: minimize } \|F(\cdot) - f(x, \cdot)\|_{\infty, Y} \text{ s.t. } x \in X, \tag{3.2}$$

where

$$\|F(\cdot) - f(x, \cdot)\|_{\infty, Y} := \max_{y \in Y} \|F(y) - f(x, y)\|.$$

This problem can be rewritten as

$$\text{CA-SIP: } \underset{(x,z) \in X \times \mathbb{R}}{\text{minimize}} z \text{ s.t. } -z \leq F(y) - f(x, y) \leq z \text{ for all } y \in Y.$$

CA-SIP is a standard semi-infinite programming problem. Note that CA-SIP is a smooth optimization problem in addition to CA, which is nonsmooth due to the maximum norm. Here, the difficulty is shifted into infinitely many inequality constraints.

Next, we formulate the problem of reverse Chebyshev approximation (RCA). Let F be a real-valued continuous function on a nonempty and compact set $Y(x) \subset \mathbb{R}^m$ that depends on a parameter $x \in X \subset \mathbb{R}^n$. Given an approximating family of functions $f(z, \cdot), z \in Z \subset \mathbb{R}^k$, and a desired precision $e(z, x)$, the aim is to find parameter vectors z and x such that the domain $Y(x)$ is as large as possible without exceeding the approximation error $e(z, x)$. This yields the problem

$$\text{RCA: } \underset{(z,x) \in Z \times X}{\text{maximize}} \text{vol}(Y(x)) \text{ s.t. } \|F(\cdot) - f(z, \cdot)\|_{\infty, Y(x)} \leq e(z, x).$$

Again, this nonsmooth optimization problem can be reformulated with semi-infinite constraints. However, we now obtain a generalized semi-infinite programming problem,

$$\text{RCA-GSIP: } \underset{(z,x) \in Z \times X}{\text{maximize}} \text{vol}(Y(x))$$

$$\text{s.t. } -e(z, x) \leq F(y) - f(z, y) \leq e(z, x) \text{ for all } y \in Y(x).$$

Design centering

In a design-centering (DC) problem, we try to maximize the size of a parameterized body $B(x)$ contained in a container set C:

$$\text{DC: } \underset{x\in\mathbb{R}^n}{\text{maximize}} \text{vol}(B(x)) \text{ s.t. } B(x) \subset C.$$

Let the container C be given by inequality constraints:

$$C = \{y \in \mathbb{R}^m \,|\, g(y) \leq 0\}.$$

Then, DC can equivalently be written as

$$\text{DC-GSIP: } \underset{x\in\mathbb{R}^n}{\text{maximize}} \text{vol}(B(x)) \text{ s.t. } g(y) \leq 0 \text{ for all } y \in B(x).$$

Robust optimization

Robust optimization deals with an a priori analysis of optimization problems depending on uncertain data (see [7]). The so-called robust counterparts of finite optimization problems fit in the context of GSIPs. Let an inequality constraint function $G(x,y)$ depend on some uncertain parameter vector y from a so-called uncertainty set $Y \subset \mathbb{R}^m$. Then the pessimistic way to deal with this constraint is to use its worst-case reformulation,

$$G(x,y) \leq 0 \text{ for all } y \in Y.$$

This inequality system is of semi-infinite type. Now let the uncertainty set Y also depend on the decision variable x. We obtain a generalized semi-infinite constraint

$$G(x,y) \leq 0 \text{ for all } y \in Y(x).$$

Next, let an objective function $F(x,y)$ depend on the unknown parameter $y \in Y(x)$. In the worst case, one has to minimize the maximal objective value; that is, one considers the problem

$$\underset{x\in\mathbb{R}^n}{\text{minimize}} \underset{y\in Y(x)}{\text{maximize}} F(x,y).$$

Hence, we are ready to write down a robust counterpart for the parametric optimization problem

$$\text{NLP: } \underset{x\in\mathbb{R}^n}{\text{minimize}} F(x,y) \text{ s.t. } G_i(x,y) \leq 0, i \in I$$

with an unknown parameter $y \in Y(x)$. The robust counterpart is

$$\text{Robust-GSIP: } \underset{(x,z)\in\mathbb{R}^n\times\mathbb{R}}{\text{minimize}} z$$

$$\text{s.t. } F(x,y) \leq z, G_i(x,y) \leq 0 \text{ for all } y \in Y(x).$$

Nonclosedness and re-entrant corners

We present two illustrative examples that show the intrinsic difficulties of GSIPs.

Example 16 (Nonclosedness of the feasible set). Let $m = 1$, $s = 1$, and the GSIP feasible set M be given by

$$M = \{x \in \mathbb{R}^n \,|\, g_0(x,y) \geq 0 \text{ for all } y \in Y(x)\}$$

with

$$Y(x) = \{y \in \mathbb{R} \,|\, g_1(x,y) \leq 0\}.$$

For $\bar{x} \in \mathbb{R}^n$, the graphs of $g_0(\bar{x},\cdot)$ and $g_0(\bar{x},\cdot)$ are depicted in Figure 18. It is clear that $\bar{x} \in M$. Let $\bar{y}_1$ and $\bar{y}_2$ be local minimizers of $g_0(\bar{x},\cdot)$ and $g_0(\bar{x},\cdot)$, respectively (see Figure 18). In a neighborhood of $\bar{x}$, we parameterize by $y_1(x)$ and $y_2(x)$ the local minimizers of $g_0(x,\cdot)$ and $g_0(x,\cdot)$ such that $y_1(\bar{x}) = \bar{y}_1$ and $y_2(\bar{x}) = \bar{y}_2$. A moment of reflection shows that locally around $\bar{x}$ the feasible set M is given as

$$M \underset{\text{loc.}}{=} \{x \,|\, g_0(x,y_1(x)) \geq 0, \, g_1(x,y_2(x)) > 0\}.$$

We conclude that M is locally nonclosed.

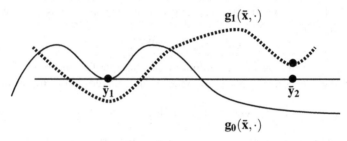

Figure 18　Graphs of $g_0(\bar{x},\cdot)$ and $g_0(\bar{x},\cdot)$ in Example 16

Example 17 (Re-entrant corners [40]). Let $n = 2$, $m = 1$, $s = 2$, and the GSIP feasible set M be given by

$$M = \left\{x \in \mathbb{R}^2 \,|\, g_0(x,y) := y \geq 0 \text{ for all } y \in Y(x)\right\}$$

with

$$Y(x) = \{y \in \mathbb{R} \,|\, g_1(x,y) := x_1 - y \leq 0, \, g_2(x,y) := y - x_2 \leq 0\}.$$

It is clear that $x \in M$ if and only if $y \geq 0$ for all $x_1 \leq y \leq x_2$. The feasible set M is depicted in Figure 19. Note that M is nonclosed and exhibits a re-entrant corner at the origin.

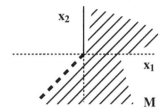

Figure 19 Feasible set M from Example 17

3.2 Structure of the feasible set

3.2.1 Closure of the feasible set and Sym-MFCQ

The feasible set M in a GSIP need not be closed. We introduce a natural constraint qualification, called the symmetric Mangasarian-Fromovitz constraint qualification (Sym-MFCQ). The Sym-MFCQ is a nontrivial extension of the well-known (extended) MFCQ for the special case of SIP and disjunctive optimization. Under the Sym-MFCQ, the closure $\overline{M}$ has an easy and natural description. As a consequence, we get a description of the interior and boundary of M. The Sym-MFCQ is shown to be generic and stable under C^1-perturbations of the defining functions. For the latter stability, consideration of the closure of M is essential. We introduce an appropriate notion of Karush-Kuhn-Tucker (KKT) points. We show that local minimizers are KKT points under the Sym-MFCQ. We refer the reader to [39] for details.

Sym-MFCQ and its consequences

Recall that the set-valued mapping $Y : \mathbb{R}^n \rightrightarrows \mathbb{R}^m$ is called locally bounded if for each $\bar{x} \in \mathbb{R}^n$ there exists a neighborhood U of $\bar{x}$ such that $\bigcup_{x \in U} Y(x)$ is bounded in $\mathbb{R}^m$.

We state the following standard assumption in the context of GSIP.

Assumption B *The mapping $Y : \mathbb{R}^n \rightrightarrows \mathbb{R}^m$ is locally bounded.*

It is well-known that the feasible set M need not be closed. Moreover, the local nonclosedness of the feasible set M is stable (e.g., [40]). Therefore, one considers the topological closure $\overline{M}$ of M instead. In [36], an explicit description of $\overline{M}$ is provided. In fact, under Assumption B and additional generic assumptions (see [36] for details), the closure of the feasible set is given by

$$\overline{M} = \left\{ x \in \mathbb{R}^n \,|\, g_0(x,y) \geq 0 \text{ for all } y \in Y^<(x) \right\}$$

with

$$Y^<(x) = \{y \in \mathbb{R}^m \mid g_k(x,y) < 0, k = 1, \ldots, s\}.$$

Using the function

$$\sigma(x,y) := \max_{0 \le k \le s} g_k(x,y),$$

it can equivalently be written as (see [37])

$$\overline{M} = \{x \in \mathbb{R}^n \mid \sigma(x,y) \ge 0 \text{ for all } y \in \mathbb{R}^m\}. \tag{3.3}$$

Note that the description of $\overline{M}$ is symmetric in the functions $g_k, k = 0, \ldots, s$. This means that the function g_0 does not play any special role.

The main goal is to present a stable and generic constraint qualification for the GSIP (see Definition 18) that provides the foregoing description of $\overline{M}$ as in (3.3). We need some auxiliary notation for its formulation.

We denote the right-hand side of (3.3) as

$$M^{\max} := \{x \in \mathbb{R}^n \mid \sigma(x,y) \ge 0 \text{ for all } y \in \mathbb{R}^m\}.$$

Note that $M^{\max}$ is a closed set due to the continuity of σ.

We set $M(\bar{x}) := \{y \in \mathbb{R}^m \mid \sigma(\bar{x},y) = 0\}$ for $\bar{x} \in M^{\max}$ and let it be empty otherwise. Note that every $y \in M(\bar{x})$ is a global minimizer of $\sigma(\bar{x}, \cdot)$ with the vanishing optimal value. Furthermore, for $y \in M(\bar{x})$ we set

$$K_0(\bar{x},y) := \{k \in \{0, \ldots, s\} \mid g_k(\bar{x},y) = 0\}.$$

Obviously, $K_0(\bar{x},y) \ne \emptyset$ for all $y \in M(\bar{x})$. Finally, $V(\bar{x},y)$ is defined as a compact convex subset of $\mathbb{R}^n$ by the following equality in $\mathbb{R}^n \times \mathbb{R}^m$:

$$V(\bar{x},y) \times \{0\} = (\mathbb{R}^n \times \{0\}) \cap \text{conv}\,\{Dg_k(\bar{x},y) \mid k \in K_0(\bar{x},y)\};\ \text{i.e.,}$$

$$V(\bar{x},y) = \left\{ \sum_{k \in K_0(\bar{x},y)} \gamma_k D_x g_k(\bar{x},y) \;\middle|\; \begin{array}{c} \sum_{k \in K_0(\bar{x},y)} \gamma_k D_y g_k(\bar{x},y) = 0, \\ \sum_{k \in K_0(\bar{x},y)} \gamma_k = 1, \\ \gamma_k \ge 0 \end{array} \right\}.$$

It is clear that $V(\bar{x},y) \ne \emptyset$ for all $y \in M(\bar{x})$. Moreover, we put

$$V(\bar{x}) := \bigcup_{y \in M(\bar{x})} V(\bar{x},y). \tag{3.4}$$

In order to indicate the dependence on the data functions $g := (g_0, \ldots, g_s)$, we write $M_g^{\max}$, $M_g(\bar{x})$, $K_0^g(\bar{x},y)$, and $V_g(\bar{x},y)$ if needed.

Definition 18 (Sym-MFCQ). Let $\bar{x} \in M^{\max}$. The symmetric Mangasarian-Fromovitz constraint qualification (Sym-MFCQ) is said to hold at $\bar{x}$ if there exists a vector $\xi \in \mathbb{R}^n$ such that for all $v \in V(\bar{x})$ it holds that

$$v \cdot \xi > 0.$$

Remark 17. It is worth mentioning that the Sym-MFCQ was indicated already in [36]. Sym-MFCQ is also connected with a constraint qualification for GSIP proposed in [72]. Indeed, it is not difficult to see that $M \subset M^{\max}$. Then, for $\bar{x} \in M$, the Sym-MFCQ coincides with the extended Mangasarian-Fromovitz constraint qualification (EMFCQ) for GSIPs as proposed in the final remarks in [72]. In addition, the Sym-MFCQ also provides a condition for the points from $M^{\max} \backslash M$. We emphasize that these points have to be regarded because of the possible nonclosedness of the feasible set M.

The following example shows that a "naive" generalization of the standard MFCQ fails here.

Example 18. With $g_1(x,y) := -2x + y$ and $g_2(x,y) := x - y$, we consider the set

$$M^{\max} := \{x \in \mathbb{R} \,|\, \max\{g_1(x,y), g_2(x,y)\} \geq 0 \text{ for all } y \in \mathbb{R}\}.$$

It is easy to see that

$$\psi(x) := \min_{y \in \mathbb{R}} \max\{g_1(x,y), g_2(x,y)\} = -\frac{1}{2}x$$

and

$$M^{\max} = \{x \in \mathbb{R} \,|\, \psi(x) \geq 0\} = (-\infty, 0].$$

For the boundary point of $M^{\max}$, $\bar{x} = 0$, we have $M(\bar{x}) = \{0\}$. With $D_x g_1(0,0) = -2$ and $D_x g_2(0,0) = 1$, the "naive" generalization of the standard MFCQ at $\bar{x}$ fails. Namely, there does not exist a real number ξ such that

$$D_x g_1(0,0)\xi > 0 \text{ and } D_x g_2(0,0)\xi > 0.$$

Nevertheless, it is easy to see that $V(0,0) = \left\{-\dfrac{1}{2}\right\}$ Hence, the Sym-MFCQ holds at 0 and $\{D_x \psi(0)\} = V(0,0)$. Moreover, the zero of ψ remains stable under small C^1-perturbations of g_1 and g_2.

The following simple reformulation of the Sym-MFCQ was indicated by one of the anonymous referees.

Proposition 7. *Let Assumption B be satisfied. The Sym-MFCQ holds at $\bar{x} \in M^{\max}$ if and only if there exists a vector $\eta \in \mathbb{R}^{n+m}$ such that for all $y \in M(\bar{x})$ and $k \in K_0(\bar{x}, y)$ it holds that*

$$Dg_k(\bar{x}, y) \cdot \eta > 0.$$

The proof of Proposition 7 and the subsequent theorems are given below.

We state the main results concerning the Sym-MFCQ and its impacts on the feasible set M.

Let $\mathscr{A}$ denote the set of problem data $(f, g_0, \ldots, g_s) \in C^1(\mathbb{R}^n) \times \left[C^1(\mathbb{R}^n \times \mathbb{R}^m)\right]^{s+1}$ such that Assumption B is satisfied. The set $\mathscr{A}$ is C_s^0-open (see [68]).

Theorem 22 (Sym-MFCQ is stable and generic). *Let $\mathscr{F}$ denote the subset of $\mathscr{A}$ consisting of those problem data $(f, g_0, \ldots, g_s)$ for which the Sym-MFCQ holds at all points $\bar{x} \in M^{max}$. Then, $\mathscr{F}$ is C_s^1-open and C_s^1-dense in $\mathscr{A}$.*

Theorem 23 (Closure theorem). *Let the Sym-MFCQ hold at all points $\bar{x} \in M^{max}$ and Assumption B be satisfied. Then, $\overline{M} = M^{max}$.*

Theorem 24 (Topological properties of M). *Let Sym-MFCQ hold at all points $\bar{x} \in M^{max}$ and Assumption B be satisfied. Then*

(i) $int(\overline{M}) = int(M)$,
(ii) $\partial \overline{M} = \partial M$.

Proofs of main results

First, we provide a local description of M^{max} that is crucial for the following.

Lemma 7 (Local description of M^{max} [37]). *Let Assumption B be satisfied. For $\bar{x} \in M^{max}$, there exist some neighborhood U of $\bar{x}$ and a nonempty compact set $V \subset \mathbb{R}^m$ such that*

$$M^{max} \cap U = \{x \in U \mid \sigma(x,y) \geq 0 \text{ for all } y \in V\} = \{x \in U \mid \psi(x) \geq 0\}$$

with the well-defined continuous function $\psi(x) := \min_{y \in V} \sigma(x,y)$.

If additionally $\psi(\bar{x}) = 0$, then $M(\bar{x}) = \{y \in V \mid \sigma(\bar{x},y) = 0\}$.

Proof. One chooses as V the closure of the bounded set $\bigcup_{x \in U} Y(x)$ for the neighborhood U of $\bar{x}$ from Assumption B. $\square$

Proof of Proposition 7

(a) Let $\eta \in \mathbb{R}^{n+m}$ be a vector such that

$$Dg_k(\bar{x},y) \cdot \eta > 0 \text{ for all } y \in M(\bar{x}) \text{ and } k \in K_0(\bar{x},y). \tag{3.5}$$

Putting $\eta = (\eta_1, \eta_2) \in \mathbb{R}^{n+m}$, we show that for all $v \in V(\bar{x})$ it holds that

$$v \cdot \eta_1 > 0.$$

Indeed, let $v = \sum_{k \in K_0(\bar{x},y)} \gamma_k D_x g_k(\bar{x},y)$ with

$$\sum_{k \in K_0(\bar{x},y)} \gamma_k D_y g_k(\bar{x},y) = 0, \quad \sum_{k \in K_0(\bar{x},y)} \gamma_k = 1, \; \gamma_k \geq 0.$$

Multiplying (3.5) by γ_k and summing up w.r.t. $k \in K_0(\bar{x},y)$, we get:

$$\sum_{k \in K_0(\bar{x},y)} \gamma_k D_x g_k(\bar{x},y) \cdot \eta_1 + \sum_{k \in K_0(\bar{x},y)} \gamma_k D_y g_k(\bar{x},y) \cdot \eta_2 = v \cdot \eta_1 > 0.$$

The latter means that the Sym-MFCQ holds at $\bar{x}$.

(b) Assume that there does not exist a vector $\eta \in \mathbb{R}^{n+m}$ such that

$$Dg_k(\bar{x},y) \cdot \eta > 0 \text{ for all } y \in M(\bar{x}) \text{ and } k \in K_0(\bar{x},y).$$

Due to the compactness of $M(\bar{x})$ (see Lemma 7), a separation argument can be used as in [110], and we obtain

$$0 \in \text{conv} \left\{ D^T g_k(\bar{x},y) \,|\, y \in M(\bar{x}), k \in K_0(\bar{x},y) \right\}.$$

Hence, there exist $y_i \in M(\bar{x})$ and $\gamma_k^i \geq 0$, $k \in K_0(\bar{x},y_i)$, $i = 1,\dots,l$ such that

$$\sum_{i=1}^{l} \sum_{k \in K_0(\bar{x},y_i)} \gamma_k^i Dg_k(\bar{x},y_i) = 0. \tag{3.6}$$

We put $\gamma^i := \sum_{k \in K_0(\bar{x},y_i)} \gamma_k^i$, $i = 1,\dots,l$. Without loss of generality, we may assume $\gamma^i > 0$ for all i. Furthermore, (3.6) can be written as

$$\sum_{i=1}^{l} \gamma^i \cdot \begin{pmatrix} \displaystyle\sum_{k \in K_0(\bar{x},y_i)} \frac{\gamma_k^i}{\gamma^i} D_x g_k(\bar{x},y_i) \\ \displaystyle\sum_{k \in K_0(\bar{x},y_i)} \frac{\gamma_k^i}{\gamma^i} D_y g_k(\bar{x},y_i) \end{pmatrix} = 0.$$

This means, in particular, that

$$\sum_{k \in K_0(\bar{x},y_i)} \frac{\gamma_k^i}{\gamma^i} D_y g_k(\bar{x},y_i) = 0, \quad \sum_{k \in K_0(\bar{x},y_i)} \frac{\gamma_k^i}{\gamma^i} = 1, \quad \frac{\gamma_k^i}{\gamma^i} \geq 0,$$

and hence $v^i := \sum_{k \in K_0(\bar{x},y_i)} \frac{\gamma_k^i}{\gamma^i} D_x g_k(\bar{x},y_i) \in V(\bar{x},y_i)$. Moreover, we conclude that

$$\sum_{i=1}^{l} \gamma^i \cdot v^i = 0 \text{ with } \gamma^i > 0, \, v_i \in V(\bar{x}).$$

This latter shows that the Sym-MFCQ does not hold at $\bar{x}$. $\square$

For the proof of Theorem 22, we need some upper-semicontinuity properties of the set-valued mappings $(x,g) \rightrightarrows M_g(x)$ and $(x,y,g) \rightrightarrows V_g(x,y)$. Recall that a set-valued mapping $\mathcal{M}$ from a topological space T into a family of all subsets of $\mathbb{R}^n$ is said to be upper-semicontinuous at $\bar{v} \in T$ if, for any open set $\mathcal{U} \subset \mathbb{R}^n$ with

$\mathcal{M}(\bar{v}) \subset \mathcal{U}$, there exists a neighborhood $\mathcal{V}$ of $\bar{v}$ such that $\mathcal{M}(v) \subset \mathcal{U}$ whenever $v \in \mathcal{V}$.

Lemma 8 (Upper-semicontinuity of $M_g(x)$ and $V_g(x,y)$). *Let Assumption B be satisfied. For $\bar{x} \in M_{\bar{g}}^{max}$ and $\bar{y} \in M_{\bar{g}}(\bar{x})$, it holds that*

a)   *the set-valued mapping $(x,g) \rightrightarrows M_g(x)$ is upper-semicontinuous at $(\bar{x},\bar{g})$ w.r.t. the topology in $\mathbb{R}^n \times \left[C^1(\mathbb{R}^n \times \mathbb{R}^m) \right]^{s+1}$ and*

b)   *the set-valued mapping $(x,y,g) \rightrightarrows V_g(x,y)$ is upper-semicontinuous at $(\bar{x},\bar{y},\bar{g})$ w.r.t. the topology in $\mathbb{R}^n \times \mathbb{R}^m \times \left[C^1(\mathbb{R}^n \times \mathbb{R}^m) \right]^{s+1}$.*

Proof. (a) We assume that $(x,g) \rightrightarrows M_g(x)$ is not upper-semicontinuous at $(\bar{x},\bar{g})$. Then, there exist an open set $\mathcal{U} \subset \mathbb{R}^m$ with $M_{\bar{g}}(\bar{x}) \subset \mathcal{U}$ and $(x^i, g^i) \in \mathbb{R}^n \times \left[C^1(\mathbb{R}^n \times \mathbb{R}^m) \right]^{s+1}$, $i \in \mathbb{N}$ such that

$$(x^i, g^i) \overset{i}{\longrightarrow} (\bar{x}, \bar{g}) \text{ and } y^i \in M_{g^i}(x^i), \, y^i \notin \mathcal{U}.$$

(Strictly speaking, we should use net convergence instead of sequential convergence in the C_s^1-topology. However, the argumentation will essentially be the same.)

Now we use the representation of $M_{\bar{g}}^{max}$ and $M_{\bar{g}}(\bar{x})$ from Lemma 7 using the neighborhoods U and V as defined there. For sufficiently large $i \in \mathbb{N}$, we have $x^i \in U$ and, moreover, $y^i \in V$. Indeed, otherwise we get $y^i \notin V$ for some subsequence, denoted again by y^i. It means that there exists $k^i \in \{1, \dots, s\}$ such that $g_{k^i}(x^i, y^i) > 0$ (see the definition of V from Lemma 7). After taking an appropriate subsequence if needed, we obtain for some $k \in \{1, \dots, s\}$

$$g_k(x^i, y^i) > 0. \tag{3.7}$$

From $y^i \in M_{g^i}(x^i)$, we also obtain

$$g_k^i(x^i, y^i) \le 0. \tag{3.8}$$

Together, (3.7) and (3.8) contradict the fact that $g^i \overset{i}{\longrightarrow} \bar{g}$ in the C_s^1-topology. Furthermore, since V is compact, we assume, w.l.o.g., that $y^i \overset{i}{\longrightarrow} \bar{y} \in V$. Thus, from $y^i \in M_{g^i}(x^i)$ (i.e. $\sigma_{g^i}(x^i, y^i) = 0$, $i \in \mathbb{N}$), it follows that

$$\sigma_{\bar{g}}(\bar{x}, \bar{y}) = 0 \text{ and } \bar{y} \in M_{\bar{g}}(\bar{x}).$$

From $M_{\bar{g}}(\bar{x}) \subset \mathcal{U}$, we obtain that $\bar{y} \in \mathcal{U}$. This contradicts the fact that $y^i \overset{i}{\longrightarrow} \bar{y}$ and $y^i \notin \mathcal{U}$.

(b) We assume that $(x,y,g) \rightrightarrows V_g(x,y)$ is not upper-semicontinuous at $(\bar{x},\bar{y},\bar{g})$. Then, there exist an open set $\mathcal{U} \subset \mathbb{R}^n$ with $V_{\bar{g}}(\bar{x},\bar{y}) \subset \mathcal{U}$ and $(x^i,y^i,g^i) \in \mathbb{R}^n \times \mathbb{R}^m \times \left[C^1(\mathbb{R}^n \times \mathbb{R}^m) \right]^{s+1}$, $i \in \mathbb{N}$ such that

$$(x^i, y^i, g^i) \overset{i}{\longrightarrow} (\bar{x}, \bar{y}, \bar{g}) \text{ and } v^i \in V_{g^i}(x^i, y^i), \, v^i \notin \mathcal{U}.$$

The latter means

$$v^i = \sum_{k \in K_0^{g^i}(x^i, y^i)} \gamma_k^i D_x g_k^i(x^i, y^i) \tag{3.9}$$

with

$$\sum_{k \in K_0^{g^i}(x^i, y^i)} \gamma_k^i D_y g_k^i(x^i, y^i) = 0, \quad \sum_{k \in K_0^{g^i}(x^i, y^i)} \gamma_k^i = 1, \ \gamma_k^i \geq 0. \tag{3.10}$$

Since $(x^i, y^i, g^i) \longrightarrow (\bar{x}, \bar{y}, \bar{g})$, we obtain $K_0^{g^i}(x^i, y^i) \subset K_0^{\bar{g}}(\bar{x}, \bar{y})$ for sufficiently large $i \in \mathbb{N}$. We enlarge the sum in (3.9 and 3.10) up to $K_0^{\bar{g}}(\bar{x}, \bar{y})$ with respective $\gamma_k^i = 0$. Since the sum of $\gamma_k^i, k \in K_0^{\bar{g}}(\bar{x}, \bar{y})$ is 1, we may assume, w.l.o.g., that $\left(\gamma_k^i, k \in K_0^{\bar{g}}(\bar{x}, \bar{y})\right) \longrightarrow \left(\bar{\gamma}_k, k \in K_0^{\bar{g}}(\bar{x}, \bar{y})\right)$. Taking $i \longrightarrow \infty$ in (3.9) and (3.10), we conclude that

$$\bar{v} \in V_{\bar{g}}(\bar{x}, \bar{y}) \text{ with } v^i \xrightarrow{i} \bar{v}.$$

From $V_{\bar{g}}(\bar{x}, \bar{y}) \subset \mathscr{U}$, we obtain that $\bar{v} \in \mathscr{U}$. This contradicts the fact that $v^i \xrightarrow{i} \bar{v}$ and $v^i \notin \mathscr{U}$. $\square$

The description of $M(\bar{x})$ in Lemma 7 easily provides the following result.

Lemma 9. *Let Assumption B be satisfied. For $\bar{x} \in M^{max}$ with $\psi(\bar{x}) = 0$, the set $V(\bar{x})$ is compact.*

We state the symmetric linear independence constraint qualification (Sym-LICQ) for the GSIP which is shown to be stronger than the Sym-MFCQ.

Definition 19 (Sym-LICQ [38]). Let $\bar{x} \in M^{max}$. The symmetric linear independence constraint qualification (Sym-LICQ) is said to hold at $\bar{x}$ if for any finite subset $\{y_1, \ldots, y_p\} \subset M(\bar{x})$ and any choice of $v_i \in V(\bar{x}, y_i), i = 1, \ldots, p$, the vectors $\{v_1, \ldots, v_p\}$ are linearly independent.

Lemma 10 (Sym-LICQ implies Sym-MFCQ). *Let Assumption B be satisfied. If the Sym-LICQ holds at $\bar{x} \in M^{max}$, then the Sym-MFCQ holds as well.*

Proof. Let the Sym-LICQ hold at $\bar{x} \in M^{max}$. Without loss of generality, we may assume that $\psi(\bar{x}) = 0$ and hence $M(\bar{x}) \neq \emptyset$ (otherwise, the Sym-MFCQ holds trivially). In particular, the Sym-LICQ implies that $M(\bar{x})$ is finite and we have

$$M(\bar{x}) = \{y_1, \ldots, y_l\}, l \in \mathbb{N}.$$

Now, we assume that the Sym-MFCQ does not hold at $\bar{x}$. Since $V(\bar{x})$ is compact (see Lemma 9), a separation argument can be used as in [110], and we obtain

$$0 \in \text{conv}(V(\bar{x})).$$

Thus, with some finite index sets $J_i, i = 1, \ldots, l$,

$$\sum_{i=1}^{l} \sum_{v_j \in V(\bar{x},y_i),\ j \in J_i} \lambda_{i,j} v_j = 0, \quad \sum_{i,j} \lambda_{i,j} = 1, \quad \lambda_{i,j} \geq 0.$$

For $i \in \{1,\ldots,l\}$, we assume, w.l.o.g., that $\lambda_{i,j} \neq 0$ for at least one $j \in J_i$. For $i \in \{1,\ldots,l\}$, set

$$b_i := \sum_{j \in J_i} \lambda_{i,j} > 0. \tag{3.11}$$

Furthermore, we write for $v_j \in V(\bar{x},y_i)$, $j \in J_i$,

$$v_j = \sum_{k \in K_0(\bar{x},y_i)} \mu_{i,j}^k D_x g_k(\bar{x},y_i)$$

with

$$\sum_{k \in K_0(\bar{x},y_i)} \mu_{i,j}^k D_y g_k(\bar{x},y_i) = 0, \quad \sum_{k \in K_0(\bar{x},y_i)} \mu_{i,j}^k = 1, \quad \mu_{i,j}^k \geq 0.$$

Setting $a_i^k := \displaystyle\sum_{v_j \in V(\bar{x},y_i),\ j \in J_i} \frac{1}{b_i} \lambda_{i,j} \mu_{i,j}^k \geq 0$ and $\tilde{v}_i := \displaystyle\sum_{k \in K_0(\bar{x},y_i)} a_i^k D_x g_k(\bar{x},y_i)$, we obtain

$\tilde{v}_i \in V(\bar{x},y_i)$ and $\displaystyle\sum_{i=1}^{l} b_i \tilde{v}_i = 0$. Thus, the Sym-LICQ at $\bar{x}$ implies that $b_i = 0$ for all $i \in \{1,\ldots,l\}$. This provides a contradiction to (3.11). $\square$

Proof of Theorem 22

(a) We prove that $\mathscr{F}$ is C_s^1-open in $\mathscr{A}$.

(1) **Local argument.** First we prove the following assertion. Let the Sym-MFCQ hold at $\bar{x} \in M_{\bar{g}}^{\max}$ (with the vector $\xi \in \mathbb{R}^n$ as in Definition 18). Then, there exist an open neighborhood $U_{\bar{x}}$ of $\bar{x}$ and a C^1-neighborhood $W_{\bar{g}}$ of $\bar{g}$ such that

$$v^T \xi > 0 \text{ for all } v \in V_g(x,y),\ g \in W_{\bar{g}},\ x \in U_{\bar{x}} \cap M_g^{\max},\ y \in M_g(x). \tag{3.12}$$

The assertion (3.12) is of local nature. Therefore, we may use the representations of $M_{\bar{g}}^{\max}$ and $M_{\bar{g}}(\bar{x})$ from Lemma 7 with the neighborhoods U and V as defined there.

Due to the compactness of $V(\bar{x})$ (see Lemma 9), the Sym-MFCQ at $\bar{x}$ provides the existence of an open set $\tilde{V}$ such that

$$V(\bar{x},y) \subset \tilde{V} \text{ for all } y \in M_{\bar{g}}(\bar{x})$$

and

$$v^T \xi > 0 \text{ for all } v \in \tilde{V}.$$

We apply to $\tilde{V}$ the upper-semicontinuity property of $(x,y,g) \rightrightarrows V_g(x,y)$ at $(\bar{x},\bar{y},\bar{g})$, $\bar{y} \in M_{\bar{g}}(\bar{x})$ (see Lemma 8). We obtain the existence of an open neighborhood $U_{\bar{x}}(\bar{y}) \times V_{\bar{y}} \times W_{\bar{g}}(\bar{y})$ of $(\bar{x},\bar{y},\bar{g})$ such that

$$V_g(x,y) \subset \tilde{V} \text{ for all } (x,y,g) \in U_{\bar{x}}(\bar{y}) \times V_{\bar{y}} \times W_{\bar{g}}(\bar{y}).$$

The family of sets $\{V_{\bar{y}}, \bar{y} \in M_{\bar{g}}(\bar{x})\}$ constitutes an open covering of a compact set $M_{\bar{g}}(\bar{x})$ (see Lemma 7). Taking its finite subcovering $\{V_{\bar{y}_i}, i = 1, \ldots, p\}$, we obtain

$$M_{\bar{g}}(\bar{x}) \subset \widetilde{M} \text{ with } \widetilde{M} := \bigcup_{i=1}^{p} V_{\bar{y}_i}.$$

We apply to $\widetilde{M}$ the upper-semicontinuity property of $(x,g) \rightrightarrows M_g(x)$ at $(\bar{x}, \bar{g})$ (see Lemma 8). We obtain the existence of an open neighborhood $\widetilde{U}_{\bar{x}} \times W_{\bar{g}}$ of $(\bar{x}, \bar{g})$ such that

$$M_g(x) \subset \widetilde{M} \text{ for all } (x,g) \in \widetilde{U}_{\bar{x}} \times W_{\bar{g}}.$$

Finally, we define the open neighborhoods of $\bar{x}$ and $\bar{g}$, respectively:

$$U_{\bar{x}} := \bigcap_{i=1}^{p} U_{\bar{x}}(\bar{y}_i) \cap \widetilde{U}_{\bar{x}}, \, W_{\bar{g}} := \bigcap_{i=1}^{p} W_{\bar{g}}(\bar{y}_i) \cap \widetilde{W}_{\bar{g}}.$$

These neighborhoods fit the local argument (3.12).

(2) **Global argument.** The global argument is standard. From the local argument, we define a global vector field $\xi(\cdot)$ via a C^{∞}-partition of unity. Since the set of Sym-MFCQ vectors is convex, this vector field $\xi(\cdot)$ fits the Sym-MFCQ under C^1-perturbations of defining functions in C_s^1-topology.

(b) We prove that $\mathscr{F}$ is C_s^1-generic in $C^1(\mathbb{R}^n) \times [C^1(\mathbb{R}^n \times \mathbb{R}^m)]^{s+1}$. This implies that $\mathscr{F}$ is C_s^1-dense and hence $\mathscr{F} \cap \mathscr{A}$ C_s^1-dense in $\mathscr{A}$.

Let $\mathscr{G}$ denote the subset of $\mathscr{A}$ consisting of those problem data $(f, g_0, \ldots, g_s)$ for which the Sym-LICQ holds at all points $\bar{x} \in M^{\max}$. From Lemma 10, it suffices to prove that $\mathscr{G}$ is C_s^1-generic. This result is given in [36]. We briefly recapitulate its proof for the sake of completeness.

The proof is based on an application of the jet transversality theorem; for details, see, for example, [63] and Section B.2. For $p \in \mathbb{N}$, $K_i \subset \{0, \ldots, s\}$ and $r_i \in \mathbb{N}$, $r_i \leq m$, $i = 1, \ldots, p$, we consider the set Γ of $(x, y_1, \ldots, y_p, v_1, \ldots, v_p,)$ such that the following conditions are satisfied:

(i) $x \in \mathbb{R}^n$, $y_i \in \mathbb{R}^m$ (pairwise different), $v^i \in \mathbb{R}^n$.
(ii) $K_i = K_0(x, y_i)$.
(iii)   span $\{D_y g_k(x, y_i), k \in K_i\}$ has dimension r_i.
(iv) $(v_i, 0) \in$ span $\{Dg_k(x, y_i), k \in K_i\}$.
(v) $\|v_i\| = 1$.
(vi)   The vectors $\{v_1, \ldots, v_p\}$ are linearly dependent.

Now, it suffices to prove that Γ is generically empty. In fact, the available degrees of freedom of the variables involved in Γ are $n + pm + pn$ (see (i)). The loss of freedom, caused by the independent equations given in (ii)–(vi), can be counted as follows. For each $i \in \{1, \ldots, p\}$, condition (ii) generates a loss of $|K_i|$ degrees of freedom, and condition (iii) reduces the freedom by $(|K_i| - r_i)(m - r_i) \geq (m - r_i)$ degrees. Since the subspace of $\mathbb{R}^n$ formed by those vectors v_i, satisfying (iv), has at most dimension $|K_i| - r_i$, (iv) causes the loss of at least $n - |K_i| + r_i$ degrees of freedom. Condition (v) reduces the freedom by 1 degree per index i. Condition (vi) defines the loss of freedom by at least $n - p + 1$. From this, we claim that the loss of freedom is at least

$$p\left(|K_i| + (m - r_i) + (n - |K_i| + r_i) + 1\right) + n - p + 1 = pm + pn + n + 1$$

degrees. This exceeds the total available freedom $n + pm + pn$ by 1. By virtue of the jet transversality theorem, generically the set Γ must be empty. $\square$

Lemma 11. *Let the Sym-MFCQ hold at* $\bar{x} \in M^{max}$, *with the Sym-MFCQ vector* ξ *as in Definition 18. Moreover, let Assumption B be satisfied. Define the function*

$$\varphi : \begin{cases} M(\bar{x}) \longrightarrow \quad \mathbb{R}, \\ y \quad \longrightarrow \min_{v \in V(\bar{x}, y)} v^T \xi \, . \end{cases}$$

φ *is well-defined and* $\inf_{y \in M(\bar{x})} \varphi(y) > 0$.

Proof. Due to the compactness of $V(\bar{x}, y)$, φ is well-defined. Moreover, $\varphi(y) > 0, y \in M(\bar{x})$ from the Sym-MFCQ. We assume that $\inf_{y \in M(\bar{x})} \varphi(y) = 0$. Then, w.l.o.g. we may assume that there exist $y_i \stackrel{i}{\longrightarrow} \bar{y} \in M(\bar{x})$ such that $\varphi(y_i) \stackrel{i}{\longrightarrow} 0$ (recall that $M(\bar{x})$ is compact as in Lemma 7). We choose a vector $v_i \in V(\bar{x}, y_i)$ with $\varphi(y_i) = v_i^T \xi$. Using particular representations of $v_i \in V(\bar{x}, y_i)$ and taking a subsequence if needed, we may assume that $v_i \stackrel{i}{\longrightarrow} \bar{v} \in V(\bar{x}, \bar{y})$. Finally, we obtain $\bar{v}^T \xi = 0$ for $\bar{v} \in V(\bar{x}, \bar{y})$, $\bar{y} \in M(\bar{x})$. This fact contradicts the Sym-MFCQ at $\bar{x}$. $\square$

Proof of Theorem 23

Using the notation from [36], we define the following sets:

$$N^{\leq} := \{(x, y) \in \mathbb{R}^n \times \mathbb{R}^m \mid g_i(x, y) \leq 0, \, i = 0, \ldots, s\},$$

$$N := \{(x, y) \in \mathbb{R}^n \times \mathbb{R}^m \mid g_i(x, y) < 0, \, i = 0, \ldots, s\},$$

$$N^M := \{(x, y) \in \mathbb{R}^n \times \mathbb{R}^m \mid g_0(x, y) < 0, \, g_i(x, y) \leq 0, \, i = 1, \ldots, s\}.$$

Let $\Pi : \mathbb{R}^n \times \mathbb{R}^m \longrightarrow \mathbb{R}^n$ denote the projection on $\mathbb{R}^n$. Obviously, it holds that

$$CM = \Pi N^M, \, C\Pi N = M^{max}.$$

Since $N \subset N^M \subset N^{\leq}$, we obtain

$$C\Pi N^{\leq} \subset M \subset C\Pi N.$$

Thus, due to the closedness of $C\Pi N$ and $C\Pi N = M^{max}$, it suffices to show that

$$\overline{C\Pi N^{\leq}} = M^{max}.$$

Equivalently, we prove that

for each $\bar{x} \in \Pi N^{\leq} \cap M^{max}$ there exists $\bar{z} \in C\Pi N^{\leq}$ arbitrarily close to $\bar{x}$.   (3.13)

Let $\bar{x} \in \Pi N^{\leq} \cap M^{\max}$. We use the local representation of $M^{\max}$ from Lemma 7 with the neighborhoods U and V as defined there. Moreover, the following representation of $C\Pi N^{\leq}$ is valid with the same neighborhoods U and V:

$$C\Pi N^{\leq} \cap U = \{x \in U \mid \sigma(x,y) > 0 \text{ for all } y \in V\}.$$

Thus, we get from $\bar{x} \in \Pi N^{\leq} \cap M^{\max}$ that $\min_{y \in V} \sigma(\bar{x},y) \leq 0$. Recall that the Sym-MFCQ holds at $\bar{x}$. With the Sym-MFCQ vector ξ as in Definition 18, we set $z_t := \bar{x} + t\xi$. For (3.13), it suffices to show that

$$\text{there exists } \varepsilon > 0 \text{ such that } z_t \in C\Pi N^{\leq} \text{ for all } t \in (0,\varepsilon). \tag{3.14}$$

Using the representation of $C\Pi N^{\leq}$ above, we consider two cases for $y \in V$.

Case 1. $y \notin M(\bar{x})$. Thus, we obtain the existence of $k \in \{0,\ldots,s\}$ such that $g_k(\bar{x},y) > 0$.

Case 2. $y \in M(\bar{x})$. We write the Taylor expansion for $g_k(\cdot,y)$, $k \in K_0(\bar{x},y)$, at $\bar{x}$:

$$g_k(z_t,y) = t \left[D_x g_k(\bar{x},y)\xi + \frac{o_k(t,y)}{t} \right]. \tag{3.15}$$

We choose a vector $v \in V(\bar{x},y)$ written as

$$v = \sum_{k \in K_0(\bar{x},y)} \gamma^k(y) D_x g_k(\bar{x},y)$$

with

$$\sum_{k \in K_0(\bar{x},y)} \gamma^k(y) D_y g_k(\bar{x},y) = 0, \quad \sum_{k \in K_0(\bar{x},y)} \gamma^k(y) = 1, \; \gamma^k(y) \geq 0.$$

Multiplying (3.15) by $\gamma_k(y)$ and summing, we obtain

$$\sum_{k \in K_0(\bar{x},y)} \gamma^k(y) g_k(z_t,y) = t \left[v^T \xi + \sum_{k \in K_0(\bar{x},y)} \gamma^k(y) \frac{o_k(t,y)}{t} \right] \geq$$

$$t \left[\min_{v \in V(\bar{x},y)} v^T \xi + \sum_{k \in K_0(\bar{x},y)} \gamma^k(y) \frac{o_k(t,y)}{t} \right] \geq$$

$$t \left[\inf_{y \in M(\bar{x})} \min_{v^T \in V(\bar{x},y)} v\xi + \sum_{k \in K_0(\bar{x},y)} \gamma^k(y) \frac{o_k(t,y)}{t} \right].$$

From Lemma 11, $\inf_{y \in M(\bar{x})} \min_{v \in V(\bar{x},y)} v^T \xi > 0$. Moreover,

$$\sum_{k \in K_0(\bar{x},y)} \gamma^k(y) \frac{o_k(t,y)}{t} \longrightarrow 0, \text{ (as } t \longrightarrow 0) \text{ uniformly on } M(\bar{x}).$$

This comes from the fact that

$$M(\bar{x}) \text{ is compact, } \sum_{k \in K_0(\bar{x},y)} \gamma^k(y) = 1, \text{ and}$$

$$\frac{o_k(t,y)}{t} = \int_0^1 [D_x g_k(\bar{x} + st\xi, y) - D_x g_k(\bar{x}, y)] \xi \, ds \text{ is continuous w.r.t. } (t,y).$$

From this, we obtain the existence of an $\varepsilon > 0$ (which is independent from $y \in M(\bar{x})$) such that for all $t \in (0, \varepsilon)$ there exists an index $k \in 0, \ldots, s$ such that $g_k(z_t, y) > 0$. Cases 1 and 2 provide (3.14) and hence the assertion. $\square$

To prove Theorem 24, we need the following description of $\partial M^{\max}$.

Lemma 12. *Let the Sym-MFCQ hold at $\bar{x} \in M^{max}$ and Assumption B be satisfied. Then,*

$$\bar{x} \in \partial M^{max} \text{ if and only if } \psi(\bar{x}) = 0,$$

where ψ is defined as in Lemma 7.

Proof. We use the local representation of $M^{\max}$ from Lemma 7 with the neighborhoods U and V as defined there,

$$M^{\max} \cap U = \{x \in U \mid \psi(x) \geq 0\},$$

where $\psi(x) := \min_{y \in V} \sigma(x,y)$.

Due to the continuity of ψ on U, if $\psi(\bar{x}) > 0$, then $\bar{x} \in \text{int}(M^{\max})$. Hence, we restrict our consideration to the case $\psi(\bar{x}) = 0$ and prove that $\bar{x} \in \partial M^{\max}$. We use the Sym-MFCQ vector ξ at $\bar{x}$ as in Definition 18. Putting $x(t) := \bar{x} - t\xi$, $t > 0$, we show that for sufficiently small $t > 0$ we have $\psi(x(t)) < 0$ and thus $x(t) \notin M^{\max}$. This implies that $\bar{x} \notin \text{int}(M^{\max})$ and, since $M^{\max}$ is a closed set, we get $\bar{x} \in M^{\max} \setminus \text{int}(M^{\max}) = \partial M^{\max}$.

Let $\bar{y} \in M(\bar{x}) \neq \emptyset$ since $\psi(\bar{x}) = 0$.

Case (a): $k \notin K_0(\bar{x}, \bar{y})$

For those k we have $g_k(\bar{x}, \bar{y}) < 0$. Because of the continuity of $g_k(\cdot, \cdot)$, it means that

$$g_k(x,y) < 0 \text{ for all } (x,y) \text{ close to } (\bar{x}, \bar{y}). \tag{3.16}$$

Case (b): $k \in K_0(\bar{x}, \bar{y})$

We set $\tilde{g}_k(t,y) := g_k(x(t), y)$ and $D\tilde{g}_k(t,y) = (-D_x g_k(\bar{x}, \bar{y})\xi, D_y g_k(\bar{x}, \bar{y}))$. We claim that there exists a vector $w \in \mathbb{R}^{m+1}$ such that

$$D\tilde{g}_k(0, \bar{y})w > 0 \text{ for all } k \in K_0(\bar{x}, \bar{y}). \tag{3.17}$$

Otherwise, from Gordan's Theorem, we obtain the existence of $\gamma_k, k \in K_0(\bar{x}, \bar{y})$ such that

$$\sum_{k \in K_0(\bar{x},\bar{y})} \gamma_k D_x g_k(\bar{x}, \bar{y})\xi = 0, \quad \sum_{k \in K_0(\bar{x},\bar{y})} \gamma_k D_y g_k(\bar{x}, \bar{y}) = 0, \quad \sum_{k \in K_0(\bar{x},\bar{y})} \gamma_k = 1, \gamma_k \geq 0.$$

This fact contradicts the Sym-MFCQ at $\bar{x}$. Moreover, setting $w = (w_1, w_2) \in \mathbb{R} \times \mathbb{R}^m$, we get $w_1 < 0$ in (3.17). To see this, we recall that $\bar{y} \in M(\bar{x})$. Hence, we multiply the inequalities in (3.17) by appropriate γ_k, $k \in K_0(\bar{x}, \bar{y})$ (see the definition of $V(\bar{x}, \bar{y})$) and sum w.r.t. k afterwards. The Sym-MFCQ at $\bar{x}$ insures that $w_1 < 0$. Without loss of generality, we assume $w_1 = -1$.

Furthermore, due to the continuity of $D\widetilde{g}_k(\cdot, \cdot)$, we obtain from (3.17) that

$$D\widetilde{g}_k(t, y)w > 0 \text{ for all } t \in [0, \varepsilon), \ y \in \widetilde{V}, \ k \in K_0(\bar{x}, \bar{y}), \tag{3.18}$$

where $\varepsilon > 0$ and $\widetilde{V}$ is a convex neighborhood of $\bar{y}$.

We set $y(t) := \bar{y} - tw_2$. Thus, $y(t) \in \widetilde{V}$ for sufficiently small t. Moreover,

$$(t, y(t)) = (0, \bar{y}) - tw (\text{recall } w_1 = -1).$$

We apply the mean-value theorem and get, for $k \in K_0(\bar{x}, \bar{y})$,

$$\widetilde{g}_k(t, y(t)) = \widetilde{g}_k(0, \bar{y}) - tD\widetilde{g}_k(\widetilde{t}, \widetilde{y}) \cdot w \text{ with some } \widetilde{t} \in [0, \varepsilon], \ \widetilde{y} \in \widetilde{V}.$$

From $D\widetilde{g}_k(\widetilde{t}, \widetilde{y})w > 0$ (see (3.18)) and $\widetilde{g}_k(0, \bar{y}) = g_k(\bar{x}, \bar{y}) = 0$ for $k \in K_0(\bar{x}, \bar{y})$, it holds that

$$g_k(x(t), y(t)) = \widetilde{g}_k(t, y(t)) < 0 \text{ for all } k \in K_0(\bar{x}, \bar{y}). \tag{3.19}$$

Together, (3.16) and (3.19) provide that

$$\sigma(x(t), y(t)) < 0 \text{ for arbitrarily small } t > 0$$

since $(x(t), y(t)) \overset{t \to 0}{\longrightarrow} (\bar{x}, \bar{y})$. This shows that $x(t) \notin M^{\max}$ for arbitrarily small $t > 0$. $\square$

Proof of Theorem 24

From Theorem 23, $\overline{M} = M^{\max}$.

(i) Let $\bar{x} \in \text{int}(\overline{M})$. We use the local representation of $M^{\max} = \overline{M}$ from Lemma 7 with the neighborhoods U and V as defined there,

$$\overline{M} \cap U = \{x \in U \mid \psi(x) \geq 0\}.$$

Moreover, from the proof of Theorem 23, we obtain:

$$C\Pi N^{\leq} \cap U = \{x \in U \mid \psi(x) > 0\}.$$

Lemma 12 implies that $\psi(\bar{x}) \neq 0$. Thus, $\bar{x} \in C\Pi N^{\leq}$ and, due to the continuity of ψ, there exists a neighborhood $U_{\bar{x}}$ of $\bar{x}$ such that $U_{\bar{x}} \subset C\Pi N^{\leq}$. Obviously, $C\Pi N^{\leq} \subset M$ and hence $\bar{x} \in \text{int}(M)$. Note that $M \subset \overline{M}$ implies $\text{int}(M) \subset \text{int}(\overline{M})$. This shows assertion (i).

Finally, assertion (ii) is just a consequence of assertion (i) and the fact that $\partial M = \overline{M} \backslash \text{int}(M)$. $\square$

Sym-MFCQ and KKT points

Furthermore, we shed light on the consequences of the Sym-MFCQ for GSIPs regarding local minima and Karush-Kuhn-Tucker (KKT) points. (See also [72] and [119] on the first-order optimality conditions for GSIPs).

Note that $\bar{x} \in M$ is a local minimizer of the continuous function f on M if and only if it is a local minimizer of f on $\overline{M}$. Hence, we consider the optimization problem

$$\overline{\text{GSIP}}: \quad \text{minimize } f(x) \text{ s.t. } x \in \overline{M}$$

and introduce the notion of KKT points for $\overline{\text{GSIP}}$.

The following definition is motivated by the description (3.3) of $\overline{M}$ (which is valid under the Sym-MFCQ by virtue of Theorem 23).

Definition 20 (KKT point [37]). $\bar{x} \in \overline{M}$ is called a KKT point if there exist $y_1, \ldots, y_l \in M(\bar{x})$, $v_i \in V(x, y_i)$, and $\mu_i \geq 0, i = 1, \ldots, l$ such that

$$Df(\bar{x}) = \sum_{i=1}^{l} \mu_i v_i.$$

Theorem 25 (Local minimum is a KKT point). *Let the Sym-MFCQ hold at a local minimum $\bar{x} \in \overline{M}$ for $\overline{\text{GSIP}}$ and Assumption B be satisfied. Then, $\bar{x}$ is a KKT point.*

Proof. Here, we use some ideas from [72]. Let $\bar{x} \in \overline{M}$ be a local minimum for $\overline{\text{GSIP}}$. Moreover, let the Sym-MFCQ hold at $\bar{x}$ with the Sym-MFCQ vector ξ as in Definition 18. Recall that $\overline{M} \subset M^{\max}$ and hence $\bar{x} \in M^{\max}$. By the local argument (see the proof of Theorem 22), there exists a neighborhood $\widetilde{U}$ of $\bar{x}$ such that the Sym-MFCQ holds at all $M^{\max} \cap \widetilde{U}$ with the same vector ξ. Thus, from the proof of Theorem 23, it follows that

$$\overline{M} \cap \widetilde{U} = M^{\max} \cap \widetilde{U}.$$

Now, we use the local representation of $M^{\max}$ from Lemma 7 with the neighborhoods U and V as defined there. Shrinking the neighborhood U if needed, we get

$$\overline{M} \cap U = \{x \in U \mid \sigma(x, y) \geq 0 \text{ for all } y \in V\} = \{x \in U \mid \psi(x) \geq 0\}$$

with the well-defined continuous function $\psi(x) := \min_{y \in V} \sigma(x, y)$.

Case 1: $\psi(\bar{x}) > 0$

Here, the local minimum $\bar{x}$ lies in the interior of $\overline{M} \cap U$ and hence $Df(\bar{x}) = 0$. We see that $\bar{x}$ is a KKT point as in Definition 20.

Case 2: $\psi(\bar{x}) = 0$

Then, $M(\bar{x}) \neq \emptyset$, and Lemma 7 provides the representation

$$M(\bar{x}) = \{y \in V \mid \sigma(\bar{x}, y) = 0\}.$$

We assume that

$$0 \notin \mathrm{conv}\left(\{-D^T f(\bar{x})\} \cup V(\bar{x})\right). \tag{3.20}$$

Due to the compactness of $V(\bar{x})$ (see Lemma 9), a separation argument can be used as in [110]. From (3.20), we obtain the existence of a vector $\xi \in \mathbb{R}^n$ such that

$$D^T f(\bar{x})\xi < 0$$

and

$$v^T \xi > 0 \text{ for all } v \in V(\bar{x}). \tag{3.21}$$

Since $\bar{x}$ is a local minimum of f on $\overline{M} \cap U$ and $D^T f(\bar{x})\xi < 0$, there exists $\varepsilon > 0$ such that for all $t \in (0, \varepsilon]$

$$f(\bar{x}+t\xi) < f(\bar{x}) \text{ and } \bar{x}+t\xi \notin \overline{M}.$$

We choose a sequence $(t_i)_{i \in \mathbb{N}} \subset (0, \varepsilon)$ with $t_i \xrightarrow{i} 0$ and set $x_i := \bar{x}+t_i\xi$, $i \in \mathbb{N}$, with $x_i \notin \overline{M}$. Hence, there exist $y_i \in V$, $i \in \mathbb{N}$ such that

$$\sigma(x_i, y_i) < 0 \text{ for all } i \in \mathbb{N}. \tag{3.22}$$

Without loss of generality, we assume that $y_i \xrightarrow{i} \bar{y} \in V$ and thus

$$\sigma(\bar{x}, \bar{y}) \le 0. \tag{3.23}$$

This, together with $\bar{x} \in M^{\mathrm{max}}$, means that $\sigma(\bar{x}, \bar{y}) = 0$ and $\bar{y} \in M(\bar{x})$.

To produce a contradiction with $\bar{x} \in M^{\mathrm{max}}$, we show that there exists $\widetilde{y} \in \mathbb{R}^m$ such that $\sigma(\bar{x}, \widetilde{y}) < 0$.

Case (a): $k \notin K_0(\bar{x}, \bar{y})$
From (3.23), $g_k(\bar{x}, \bar{y}) < 0$. Because of the continuity of $g_k(\bar{x}, \cdot)$, this means that

$$g_k(\bar{x}, y) < 0 \text{ for all } y \text{ close to } \bar{y}. \tag{3.24}$$

Case (b): $k \in K_0(\bar{x}, \bar{y})$
We set $\widetilde{g}_k(t, y) := g_k(\bar{x}+t\xi, y)$ and $D\widetilde{g}_k(t, y) = (D_x g_k(\bar{x}, \bar{y})\xi, D_y g_k(\bar{x}, \bar{y}))$. We claim that there exists a vector $w \in \mathbb{R}^{m+1}$ such that

$$D\widetilde{g}_k(0, \bar{y})w > 0 \text{ for all } k \in K_0(\bar{x}, \bar{y}). \tag{3.25}$$

Otherwise, from the Farkas lemma, we obtain the existence of γ_k, $k \in K_0(\bar{x}, \bar{y})$ such that

$$\sum_{k \in K_0(\bar{x}, \bar{y})} \gamma_k D_x g_k(\bar{x}, \bar{y})\xi = 0, \quad \sum_{k \in K_0(\bar{x}, \bar{y})} \gamma_k D_y g_k(\bar{x}, \bar{y}) = 0, \quad \sum_{k \in K_0(\bar{x}, \bar{y})} \gamma_k = 1, \ \gamma_k \ge 0.$$

This fact contradicts (3.21). Moreover, setting $w = (w_1, w_2) \in \mathbb{R} \times \mathbb{R}^m$, we get $w_1 > 0$ in (3.25). To see this, we recall that $\bar{y} \in M(\bar{x})$. Hence, we multiply the inequalities in (3.25) by appropriate γ_k, $k \in K_0(\bar{x}, \bar{y})$ (see the definition of $V(\bar{x}, \bar{y})$) and sum w.r.t. k

afterwards. The inequality in (3.21) ensures that $w_1 > 0$. Without loss of generality, we assume $w_1 = 1$.

Furthermore, due to the continuity of $D\widetilde{g}_k(\cdot, \cdot)$, we obtain from (3.25)

$$D\widetilde{g}_k(t, y)w > 0 \text{ for all } t \in [0, \widetilde{\varepsilon}), y \in \widetilde{V}, k \in K_0(\bar{x}, \bar{y}), \tag{3.26}$$

where $\widetilde{V}$ is a convex neighborhood of $\bar{y}$.

We set $\widetilde{y}_i := y_i - t_i w_2$, $i \in \mathbb{N}$. Thus, $\widetilde{y}_i \in \widetilde{V}$ and $y_i \in \widetilde{V}$, for sufficiently large i, because $y_i \xrightarrow{i} \bar{y}$ and $t_i \xrightarrow{i} 0$. Moreover,

$$(t_i, y_i) = (0, \widetilde{y}_i) + t_i w (\text{recall } w_1 = 1).$$

We apply the mean-value theorem and get, for $k \in K_0(\bar{x}, \bar{y})$,

$$\widetilde{g}_k(t_i, y_i) - \widetilde{g}_k(0, \widetilde{y}_i) = t_i D\widetilde{g}_k(\widetilde{t}, \widetilde{y}) \cdot w \text{ with some } \widetilde{t} \in [0, \widetilde{\varepsilon}], \widetilde{y} \in \widetilde{V}.$$

From $D\widetilde{g}_k(\widetilde{t}, \widetilde{y})w > 0$ (see (3.26)) and $\widetilde{g}_k(t_i, y_i) = g_k(x_i, y_i) < 0$ (see 3.22), it holds that

$$g_k(\bar{x}, \widetilde{y}_i) = \widetilde{g}_k(0, \widetilde{y}_i) < 0 \text{ for all } k \in K_0(\bar{x}, \bar{y}). \tag{3.27}$$

Together, (3.24) and (3.27) provide a contradiction with $\bar{x} \in \overline{M}$ since $\widetilde{y}_i \xrightarrow{i} \bar{y}$.

Finally, we obtain that (3.20) is not valid. Hence, there exist $y_1, \ldots, y_l \in M(\bar{x})$, $v_i \in V(x, y_i)$, $\mu_i \geq 0, i = 1, \ldots, l$, $\mu \geq 0$ (not all vanishing) such that

$$\mu Df(\bar{x}) = \sum_{i=1}^{l} \mu_i v_i. \tag{3.28}$$

Assume that $\mu = 0$. Then, multiplying (3.28) by a Sym-MFCQ vector ξ, we obtain

$$0 = \sum_{i=1}^{l} \mu_i v_i^T \xi.$$

Since $v_i^T \xi > 0$, we get $\mu_i = 0$, $i = 1, \ldots, l$. This contradiction shows that $\bar{x}$ is a KKT point. $\square$

The following corollary is an easy consequence of Theorem 25.

Corollary 7. *Let the Sym-MFCQ hold at a local minimum $\bar{x} \in M$ for the GSIP and Assumption B be satisfied. Then, $\bar{x}$ is a KKT point.*

Sym-MFCQ in SIP and disjunctive optimization

We consider the special case of a standard SIP, characterized by a constant set $Y := Y(x)$,

$$\text{SIP: } \quad \text{minimize } f(x) \text{ s.t. } x \in M, \tag{3.29}$$

with

$$M := \{x \in \mathbb{R}^n \,|\, g_0(x,y) \geq 0 \text{ for all } y \in Y\},$$

and a compact set

$$Y := \{y \in \mathbb{R}^m \,|\, g_k(y) \leq 0, k = 1, \ldots, s\}.$$

We recall the well-known extended Mangasarian-Fromovitz constraint qualification (EMFCQ) for the SIP.

Definition 21 (EMFCQ for SIP). Let M be given as in (3.29) and $\bar{x} \in M$. The extended Mangasarian-Fromovitz constraint qualification (EMFCQ) is said to hold at $\bar{x}$ if there exists a vector $\xi \in \mathbb{R}^n$ such that it holds that

$$D_x g_0(\bar{x},y) \cdot \xi > 0 \text{ for all } y \in E_{g_0}(\bar{x}) := \{y \in Y \,|\, g_0(\bar{x},y) = 0\}.$$

The following lemma clarifies the relationship between the EMFCQ and Sym-MFCQ in the case of the SIP.

Lemma 13 (Sym-MFCQ vs. EMFCQ in SIP). *Let M be given as in (3.29) and $\bar{x} \in M$. Then, the Sym-MFCQ holds at $\bar{x}$ if and only if both of the following conditions are fulfilled:*

(i)   EMFCQ holds at $\bar{x}$.
(ii) The standard MFCQ holds for Y at all $y \in E_{g_0}(\bar{x})$.

Proof. It is easy to see that $M(\bar{x}) = E_{g_0}(\bar{x})$. We obtain for $y \in E_{g_0}(\bar{x})$ and $v \in V(\bar{x},y)$ that

$$v = \gamma_0 D_x g_0(\bar{x},y) \text{ with}$$

$$\gamma_0 D_y g_0(\bar{x},y) + \sum_{k \in K(\bar{x},y)} \gamma_k D_y g_k(y) = 0, \gamma_0 + \sum_{k \in K(\bar{x},y)} \gamma_k = 1, \gamma_k \geq 0.$$

If $\gamma_0 = 0$, then the standard MFCQ is violated for Y at y. Conversely, if the standard MFCQ is violated for Y at $y \in E_{g_0}(\bar{x})$, then $0 \in V(\bar{x},y)$. The proof follows directly with the aid of these considerations.$\square$

Example 19 ($M \neq M^{max}$ in SIP). Lemma 13 shows that in the case of the SIP the Sym-MFCQ incorporates not only the usual EMFCQ but also the standard MFCQ for Y. The lack of the latter is closely related to the fact that M, being closed, need not equal M^{max}, even under the EMFCQ.

We consider the following example of an SIP from [36]:

$$n = m = 1, s = 1, g_0(x,y) := x - y \text{ and } g_1(y) := y(y-1)^2.$$

It can be easily seen that $Y = (-\infty, 0] \cup \{1\}$ and therefore $M = [1, \infty)$. Moreover, $M^{max} = [0, \infty]$. Setting $\bar{x} := 1$, we have $E_{g_0}(\bar{x}) = M(\bar{x}) = \{1\}$, and the EMFCQ holds at $\bar{x}$. Nevertheless, $M \neq M^{max}$. This is due to the violation of the standard MFCQ for Y at 1 and hence the violation of the Sym-MFCQ.

In some situations, the GSIP can be equivalently rewritten as a so-called disjunctive optimization problem (DISJ) (see [71]),

$$\text{DISJ:}\quad \text{minimize } f(x) \text{ s.t. } x \in M_{\text{disj}}, \tag{3.30}$$

with

$$M_{\text{disj}} := \left\{ x \in \mathbb{R}^n \,\Big|\, \min_{j \in J} \left\{ \max_{v_j \in J^j} g_{v_j}^j(x) \right\} \geq 0 \right\}$$

and

$$J := \{1, \ldots, s\}, J^j = \{1, \ldots, k_j\}, k_j \geq 1, j \in J.$$

The Mangasarian-Fromovitz constraint qualification (MFCQ) for DISJ is defined as follows.

Definition 22 (MFCQ for DISJ). Let M_{disj} be given as in (3.30) and $\bar{x} \in M_{\text{disj}}$. The Mangasarian-Fromovitz constraint qualification (MFCQ) is said to hold at $\bar{x}$ if there exists a vector $\xi \in \mathbb{R}^n$ such that it holds that

$$Dg_{v_j}^j(\bar{x}) \cdot \xi > 0 \text{ for all } v_j \in J_0^j(\bar{x}), j \in J.$$

Here, $J_0^j(\bar{x}) := \left\{ \tilde{v} \in J^j \,|\, g_{\tilde{v}}^j = \max_{v_j \in J^j} g_{v_j}^j = 0 \right\}.$

Remark 18 (Sym-MFCQ and MFCQ for DISJ). We compare M_{disj} and the local representation of M^{max} (see Lemma 7). The only difference is that in the description of M_{disj} the minimum of finitely many maximum functions is taken over a discrete set, whereas in the description of M^{max} it is minimized over a subset of $\mathbb{R}^m$. This new issue leads to a certain modification of the MFCQ for DISJ and results in the Sym-MFCQ. In fact, the derivatives of defining functions w.r.t. y-coordinates play an important role in the Sym-MFCQ and the whole analysis above.

3.2.2 Feasible set as a Lipschitz manifold

We examine the topological structure of $\overline{M}$, the closure of the GSIP feasible set. For that, we set

$$M^{\text{max}} = \left\{ x \in \mathbb{R}^n \,\Big|\, \max_{0 \leq k \leq s} g_k(x, y) \geq 0 \text{ for all } y \in \mathbb{R}^m \right\}.$$

Recall that M^{max} is proven to be the topological closure of the GSIP feasible set under Assumption B and the Sym-MFCQ (see Theorem 23). Hence, we concentrate on the upper-level set M^{max} given by a min-max function φ,

$$M^{\text{max}} = \{ x \in \mathbb{R}^n \,|\, \varphi(x) \geq 0 \},$$

where $\varphi : \mathbb{R}^n \to \mathbb{R} \cup \{-\infty\}$ is defined as

$$\varphi(x) := \inf_{y \in \mathbb{R}^m} \max_{0 \le k \le s} g_k(x,y).$$

We establish assumptions (compactness condition CC and the Sym-MFCQ) that guarantee that

$$\partial M^{\max} = \{x \in \mathbb{R}^n \mid \varphi(x) = 0\}$$

and, moreover, that $\partial M^{\max}$ is a Lipschitz manifold of dimension $n - 1$.

The compactness condition is shown to be stable under C^0-perturbations of the defining functions of φ. The Sym-MFCQ can be seen as a constraint qualification in terms of Clarke's subdifferential of the min-max function φ. Finally, we conclude that generically the closure of the GSIP feasible set is a Lipschitz manifold (with boundary).

Compactness condition

We define the compactness condition and describe its impacts.

Definition 23 (Compactness condition CC). We say that the compactness condition CC is fulfilled, if for all sequences $(x_k, y_k)_{k \in \mathbb{N}} \subset \mathbb{R}^n \times \mathbb{R}^m$ with

$$\begin{cases} x_k \to x \in \mathbb{R}^n, k \to \infty, \\ \text{either } \sigma(x_k, y_k) \to a, k \to \infty, \text{ and } a \le \varphi(x) \\ \text{or } \sigma(x_k, y_k) \to -\infty, k \to \infty, \end{cases}$$

the sequence $(y_k)_{k \in \mathbb{N}}$ contains a convergent subsequence.

An essential implication of condition CC is that we have a local description of the function φ around a given point $\bar{x} \in \mathbb{R}^n$,

$$\varphi(x) = \min_{y \in W} \sigma(x,y) \quad \text{for all } x \in U_{\bar{x}}.$$

Here, $W \subset \mathbb{R}^m$ is a compact set and $U_{\bar{x}} \subseteq \mathbb{R}^n$ is an open neighborhood of $\bar{x}$. This local description can be obtained by another, slightly weaker assumption which we will refer to as condition C^*. However, condition C^* is not stable w.r.t. C^0-perturbations of the defining functions. A counterexample then motivates the consideration of the more restrictive condition CC. It turns out that condition CC is stable and, moreover, implies condition C^*.

Definition 24 (Condition C^*). We say that condition C^* is fulfilled if

(C1) for all $x \in \mathbb{R}^n$ and sequences $(y_k)_{k \in \mathbb{N}} \subset \mathbb{R}^m$ with $\sigma(x, y_k) \to \varphi(x), k \to \infty$, there exists a convergent subsequence of $(y_k)_{k \in \mathbb{N}}$ and

(C2) the mapping $x \rightrightarrows M(x)$ is locally bounded (i.e., for all $\bar{x} \in \mathbb{R}^n$ there exists an open neighborhood $U_{\bar{x}} \subseteq \mathbb{R}^n$ of $\bar{x}$ such that $\bigcup_{x \in U_{\bar{x}}} M(x)$ is bounded.

Note that (C1) is a kind of Palais-Smale condition. Together with the standard assumption (C2), it implies the desired local description of φ.

Lemma 14 (Condition C* implies the local description of φ). *Let condition C* be fulfilled and let $\bar{x} \in \mathbb{R}^n$. Then, there exists an open neighborhood $U_{\bar{x}} \subseteq \mathbb{R}^n$ of $\bar{x}$ and a compact subset $W \subset \mathbb{R}^m$ such that*

$$\varphi(x) = \min_{y \in W} \sigma(x,y) \quad \text{for all } x \in U_{\bar{x}}.$$

Proof. Take the neighborhood $U_{\bar{x}} \subseteq \mathbb{R}^n$ of $\bar{x}$ from property (C2) of condition C*. Then

$$W := \overline{\bigcup_{x \in U_{\bar{x}}} M(x)}$$

is obviously a compact set. Now, let $(y_k)_{k \in \mathbb{N}}$ be a minimizing sequence for $\sigma(x, \cdot)$ with $x \in U_{\bar{x}}$,

$$\sigma(x,y_k) \to \varphi(x), \quad k \to \infty.$$

(C1) implies the existence of a subsequence $(y_{k_l})_{l \in \mathbb{N}}$ of $(y_k)_{k \in \mathbb{N}}$ with

$$y_{k_l} \to \bar{y} \in \mathbb{R}^m, \quad l \to \infty.$$

σ is continuous, so we have

$$\sigma(x,y_{k_l}) \to \sigma(x,\bar{y}), \quad l \to \infty.$$

Since $(y_{k_l})_{l \in \mathbb{N}}$ is a minimizing sequence, it holds that

$$\varphi(x) = \sigma(x,\bar{y}).$$

By definition, $\bar{y} \in M(x) \subseteq W.\square$

We give two examples showing that (C1) and (C2) are independent.

Example 20 (C1 $\not\Rightarrow$ C2). Take $n = m = 1$, $s = 0$, and let $\eta \in C^\infty(\mathbb{R})$ be the smooth function with

$$\eta(y) := \begin{cases} \exp\left(\frac{1}{y^2-1}\right), & |y| < 1 \\ 0, & |y| \geq 1 \end{cases}.$$

Now, define

$$g_0(x,y) := \begin{cases} -\eta(y), & x \leq 0 \\ -\eta(y) - \eta(y - \frac{1}{x}), & x > 0 \end{cases}.$$

Note that g_0 is differentiable. (C1) is fulfilled since for $x \in \mathbb{R}$ fixed, clearly $\varphi(x) < 0$ and there are at most two compact sets containing all $y \in \mathbb{R}$ with $g_0(x,y) < 0$. But (C2) is not fulfilled at $\bar{x} = 0$ since for $x > 0$ the minimizers that are induced by the term $\eta(y - \frac{1}{x})$ are arbitrarily far away from 0.

Example 21 (C2 $\not\Rightarrow$ C1). Take $n = m = 1$, $s = 0$ and $g_0(x,y) := e^{-y^2}$. Then $\varphi(x) = 0$ for all $x \in \mathbb{R}$ and, obviously, $M(x) = \emptyset$ for all $x \in \mathbb{R}$, so (C2) is trivially fulfilled. But for $\bar{x} \in \mathbb{R}$ fixed and a sequence $(y_k)_{k \in \mathbb{N}} \subset \mathbb{R}$ with $\sigma(\bar{x}, y_k) \to \varphi(\bar{x}) = 0, k \to \infty$, we have that $\|y_k\| \to \infty, k \to \infty$, so (C1) is not fulfilled.

We now give an example showing that condition C^* is not stable.

Example 22 (Condition C^* is not stable). We set $\bar{g}_0, \bar{g}_1 \in C^1(\mathbb{R}^2)$:

$$\bar{g}_0 := \begin{cases} x(y+1)^2 + 1, & y \leq -1 \\ 1 - \frac{\eta(y)}{\eta(0)}, & |y| < 1 \\ x(y-1)^2 + 1, & y \geq 1 \end{cases}, \quad \bar{g}_1 := \begin{cases} \bar{g}_0(x,y) - 1 + x^2, & |y| < 1 \\ x^2, & |y| \geq 1 \end{cases},$$

where η is the smooth function from Example 20. For x near 0, we get the following representation for $\sigma(x, y)$:

$$\sigma(x,y) = \begin{cases} \bar{g}_0(x,y), & x \geq 0 \text{ or } x < 0, |y| < 1, \\ \bar{g}_1(x,y), & x < 0, |y| \geq 1 + \sqrt{\frac{x^2-1}{x}}. \end{cases}$$

Since $y = 0$ is the unique minimizer of $\sigma(x, \cdot)$ for all x near 0, we clearly see that condition C^* is fulfilled.

Now, suppose that condition C^* is stable w.r.t. the C_s^0-topology. Then, there exists an open neighborhood $\mathcal{O} \subseteq C^0(\mathbb{R}^2)^2$ of $(\bar{g}_0, \bar{g}_1)$ that condition C^* holds for all $(g_0, g_1) \in \mathcal{O}$. Now, fix such an open neighborhood $\mathcal{O}$. We will construct a pair $(g_0, g_1) \in \mathcal{O}$ that does not fulfill condition C^*. We set

$$g_0(x,y) := \bar{g}_0(x,y) + C \cdot \eta\left(\frac{x}{\varepsilon_x}\right) \eta\left(\frac{y}{\varepsilon_y}\right), \quad C, \varepsilon_x, \varepsilon_y > 0,$$

and

$$g_1(x,y) := \bar{g}_1(x,y).$$

We now choose $C, \varepsilon_x, \varepsilon_y$ sufficiently small that $(g_0, g_1) \in \mathcal{O}$.

Then, there exists $g_{\min} \in \mathbb{R}$ and $R > 0$ with

$$g_0(x,y) \geq g_{\min} > 0 \quad \text{for all } (x,y) \in B_R(0) \times B_1(0).$$

Now, we can find an $x < 0$ with $|x| < R$ and $\varphi_{(g_0,g_1)}(x) < g_{\min}$. Here, the minimum is attained by $\sigma(x, \cdot)$ for $|y| \geq 1 + \sqrt{\frac{x^2-1}{x}}$ and it is produced by g_1 (see Figure 20). This is a contradiction to (C2).

Note that for a given $x < 0$ near 0 the existence of the minima y with $|y| \geq 1 + \sqrt{\frac{x^2-1}{x}}$ motivates Definition 23.

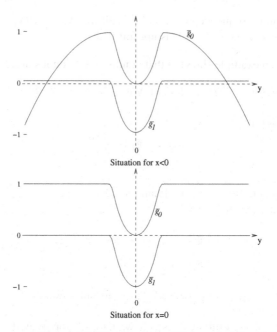

Figure 20 Illustration of Example 22

Lemma 15. *Condition CC implies condition C^*.*

Proof. Let condition CC be fulfilled. Then, trivially, (C1) holds since all minimizing sequences in the definition of (C1) are admissible sequences in definition of condition CC, and therefore the compactness of $(y_k)_{k \in \mathbb{N}}$ is implied.

We assume that (C2) does not hold. Then, for fixed $\bar{x} \in \mathbb{R}^n$ we have that for all open neighborhoods $U_{\bar{x}} \subseteq \mathbb{R}^n$ of $\bar{x}$, the set $\bigcup_{x \in U_{\bar{x}}} M(x)$ is not bounded. Hence, there exists a sequence $(x_k, y_k)_{k \in \mathbb{N}} \subset \mathbb{R}^n \times \mathbb{R}^m$ with

$$\begin{cases} x_k \to \bar{x}, k \to \infty, \\ \sigma(x_k, y_k) = \varphi(x_k) \text{ for all } k \in \mathbb{N}, \\ \|y_k\| \to \infty, k \to \infty. \end{cases} \tag{3.31}$$

Since (C1) holds, we get the existence of a minimizer for $\sigma(\bar{x}, \cdot)$; that is there exists $\bar{y} \in \mathbb{R}^m$ with $\sigma(\bar{x}, \bar{y}) = \varphi(\bar{x})$.

It holds that

$$\sigma(x_k, y_k) = \varphi(x_k) \leq \sigma(x_k, \bar{y}) \quad \text{for all } k \in \mathbb{N}.$$

Since σ is continuous, we get that $\sigma(x_k, \bar{y}) \to \sigma(\bar{x}, \bar{y}) = \varphi(\bar{x}), k \to \infty$.

Furthermore, either $(\sigma(x_k, y_k))_{k \in \mathbb{N}}$ is bounded and therefore there exists $a \in \mathbb{R}$ with $a \leq \varphi(\bar{x})$ and (regarding a subsequence if needed)

$$\sigma(x_k, y_k) \to a, \quad k \to \infty,$$

or

$$\sigma(x_k, y_k) \to -\infty, \quad k \to \infty.$$

In both cases, $(x_k, y_k)_{k \in \mathbb{N}}$ is an admissible sequence in the definition of condition CC. Together with the fact that $\|y_k\| \to \infty$, $k \to \infty$, we have a contradiction to condition CC.$\square$

We will now prove that condition CC is in fact stable under C^0-perturbations. We set

$$\mathscr{C} := \{g = (g_0, \ldots, g_s) \in C^1(\mathbb{R}^n \times \mathbb{R}^m)^{s+1} \mid \text{condition CC is fulfilled for } g\},$$

and from now on the notations $\varphi_g(x)$, $M_g(x)$, and $\sigma_g(x, y)$ indicate the dependence on the defining functions $g = (g_0, \ldots, g_s)$.

Theorem 26. *The set $\mathscr{C}$ is C_s^0-open.*

Proof. Let $\bar{g} \in \mathscr{C}$. We show that there exists an open neighborhood $\mathscr{V}_{\bar{g}} \subseteq C^0(\mathbb{R}^n \times \mathbb{R}^m)^{s+1}$ of $\bar{g}$ such that $\mathscr{V}_{\bar{g}} \subseteq \mathscr{C}$.

The proof consists of a local part and a globalization step. For the local part let $\bar{x} \in \mathbb{R}^n$ be fixed. We show that there exist open neighborhoods $U_{\bar{x}} \subseteq \mathbb{R}^n$ of $\bar{x}$ and $\mathscr{U}_{\bar{g}} \subseteq C^0(\mathbb{R}^n \times \mathbb{R}^m)^{s+1}$ of $\bar{g}$ such that

$$\text{condition CC holds at } (x, g) \quad \text{for all } (x, g) \in U_{\bar{x}} \times \mathscr{U}_{\bar{g}}. \tag{3.32}$$

Here, we say that condition CC holds at (x, g) for a pair (x, g) if the sequence $(x_k)_{k \in \mathbb{N}}$ from Definition 23 converges to x and σ, φ is replaced by σ_g, φ_g.

Now, assume that (3.32) does not hold, meaning that for all open neighborhoods $U_{\bar{x}} \subseteq \mathbb{R}^n$ of $\bar{x}$ and $\mathscr{U}_{\bar{g}} \subseteq C^0(\mathbb{R}^n \times \mathbb{R}^m)^{s+1}$ of $\bar{g}$ there exists $(x, g) \in U_{\bar{x}} \times \mathscr{U}_{\bar{g}}$ such that

$$\text{condition CC fails at } (x, g). \tag{3.33}$$

The failure of condition CC at (x, g) is by definition equivalent to the existence of a sequence $(x^{(k)}, y^{(k)})_{k \in \mathbb{N}} \subset \mathbb{R}^n \times \mathbb{R}^m$ with the properties

$$\begin{cases} x^{(k)} \to x, k \to \infty, \\ \text{either } \sigma_g(x^{(k)}, y^{(k)}) \to a, k \to \infty, \text{ and } a \le \varphi_g(x) \\ \text{or } \sigma_g(x^{(k)}, y^{(k)}) \to -\infty, k \to \infty, \\ \|y^{(k)}\| \to \infty, k \to \infty. \end{cases} \tag{3.34}$$

From Lemmas 14 and 15, we get the existence of a minimizer for $\sigma(\bar{x}, \cdot)$; that is there exists $\bar{y} \in \mathbb{R}^m$ with

$$\sigma(\bar{x}, \bar{y}) = \varphi(\bar{x}). \tag{3.35}$$

We now construct a sequence $(x_n, y_n)_{n \in \mathbb{N}}$ constituting a contradiction to condition CC at $(\bar{x}, \bar{g})$.

For that, let $n \in \mathbb{N}$ be fixed. Choose an open neighborhood $U_{\bar{x}} \subseteq \mathbb{R}^n$ of $\bar{x}$ with

$$\|x - \bar{x}\| < \frac{1}{n} \text{ and } \|\sigma(x, \bar{y}) - \sigma(\bar{x}, \bar{y})\| < \frac{1}{n}, \quad \text{for all } x \in U_{\bar{x}}, \tag{3.36}$$

and an open neighborhood $\mathscr{U}_{\bar{g}} \subseteq C^0(\mathbb{R}^n \times \mathbb{R}^m)^{s+1}$ of $\bar{g}$ such that for all $g \in \mathscr{U}_{\bar{g}}$ and for all $k \in \{0, \ldots, s\}$ it holds that

$$|g_k(x,y) - \bar{g}_k(x,y)| < \frac{1}{n} \quad \text{for all } (x,y) \in \mathbb{R}^n \times \mathbb{R}^m. \tag{3.37}$$

Then (3.33) gives us a pair $(x,g) \in U_{\bar{x}} \times \mathscr{U}_{\bar{g}}$ and a sequence $(x^{(k)}, y^{(k)})_{k \in \mathbb{N}}$ with (3.34). Thus, for k sufficiently large, we can define the n-th sequence element $(x_n, y_n) := (x^{(k)}, y^{(k)})$ and get

$$\begin{cases} \|x_n - x\| < \frac{1}{n}, \\ \sigma_g(x_n, y_n) < \varphi_g(x) + \frac{1}{n}, \\ \|y_n\| > n. \end{cases} \tag{3.38}$$

From (3.36), (3.37), and (3.38), it holds that

$$\|x_n - \bar{x}\| \leq \|x_n - x\| + \|x - \bar{x}\| < \frac{2}{n}$$

and

$$\sigma(x_n, y_n) < \sigma_g(x_n, y_n) + \frac{1}{n} < \varphi_g(x) + \frac{2}{n} \leq \sigma_g(x, \bar{y}) + \frac{2}{n}$$
$$< \sigma(x, \bar{y}) + \frac{3}{n} < \sigma(\bar{x}, \bar{y}) + \frac{4}{n}$$
$$= \varphi(\bar{x}) + \frac{4}{n},$$

which implies that $(x_n, y_n)_{n \in \mathbb{N}}$ is an admissible sequence in the definition of condition CC, and together with the property $\|y_n\| > n$ we have a contradiction to condition CC at $(\bar{x}, \bar{g})$.

The globalization step is standard. From (3.32), we get a family of neighborhoods $\{U_{\bar{x}} \times \mathscr{U}_{\bar{g}}\}_{\bar{x} \in \mathbb{R}^n}$. Then there exists a locally finite C^∞-partition of unity subsequent to the covering $\{U_{\bar{x}}\}_{\bar{x} \in \mathbb{R}^n}$ that enables us to construct a positive function $\varepsilon : \mathbb{R}^n \times \mathbb{R}^m \to \mathbb{R}_+$ inducing the desired open neighborhood $\mathscr{V}_{\bar{g}} \subseteq C^0(\mathbb{R}^n \times \mathbb{R}^m)^{s+1}$ of $\bar{g}$ from the assertion (see [63], [95] for details on this procedure). $\square$

Topological properties of $M^{\max}$

We will prove that under condition CC and the Sym-MFCQ, the upper-level set $M^{\max}$ is a Lipschitz manifold with boundary (see Definition 1).

In what follows, we use a result from [13] regarding Clarke's subdifferential of a certain optimal-value function. For that, we consider the nonlinear optimization problem

$$\text{NLP:} \quad \text{minimize } F(z) \text{ s.t. } z \in N, \tag{3.39}$$

where

$$N := \{z \in \mathbb{R}^n \mid H(z) = 0, G(z) \leq 0\}$$

and $F \in C^1(\mathbb{R}^n)$, $H \in C^1(\mathbb{R}^n)^k$, $G \in C^1(\mathbb{R}^n)^l$.

Now, we define $\psi : \mathbb{R}^k \times \mathbb{R}^l \to \mathbb{R} \cup \{-\infty\}$ as the optimal-value function of the right-hand-side perturbations of (3.39):

$$\psi(v, w) := \inf\{F(z) \mid z \in \mathbb{R}^n, H(z) + v = 0, G(z) + w \leq 0\}. \tag{3.40}$$

Definition 25 (Hypothesis H; see Hypothesis 6.5.1 in [13]). We say that Hypothesis H holds if $\psi(0,0)$ is finite and there exists a compact subset $K \subset \mathbb{R}^n$ and $\varepsilon > 0$ such that for all $(v, w) \in \mathbb{R}^k \times \mathbb{R}^l$ with $\|(v, w)\| < \varepsilon$ and $\psi(v, w) < \psi(0,0) + \varepsilon$ we have that $\psi(v, w)$ is finite and has a solution in K.

The proof of the following result is given in [13].

Theorem 27 (see Theorem 6.5.2 in [13]). *Let Hypothesis H be fulfilled. Then, it holds that*

$$\partial \psi(0,0) \subseteq \overline{conv(\Delta^1(M_0) + \Delta^0(M_0))},$$

where

$$\Delta^\delta(z) := \{(\lambda, \mu) \in \mathbb{R}^k \times \mathbb{R}^l \mid D_z L(z, \delta, \lambda, \mu) = 0, \mu_j \geq 0, \mu_j G_j(z) = 0\},$$

$$L(z, \delta, \lambda, \mu) := \delta F(z) + \sum_{i=1}^k \lambda_i H_i(z) + \sum_{j=1}^l \mu_j G_j(z),$$

$$K_0(z) := \{j \in \{1, \ldots, l\} \mid G_j(z) = 0\},$$
$$M_0 := \{z \in \mathbb{R}^n \mid F(z) = \psi(0,0), H(z) = 0, G(z) \leq 0\}.$$

We apply Theorem 27 to obtain an inclusion for Clarke's subdifferential of φ. For that, we write φ as an optimal-value function of the type (3.40). Note that

$$\varphi_{g+w}(x) = \psi(x - \bar{x}, w) \quad \text{for all } x \in \mathbb{R}^n \text{ and } w \in \mathbb{R}^{s+1}, \tag{3.41}$$

where in (3.40) we define $z := (u, y, a) \in \mathbb{R}^n \times \mathbb{R}^m \times \mathbb{R}$, $F(z) := a$, $H(z) := -u$, and $G(z) := (g_0(u + \bar{x}, y) - a, \ldots, g_s(u + \bar{x}, y) - a)^T$.

To see that the identity (3.41) is in fact valid, we have to check that

$$\inf_{y \in \mathbb{R}^m} \sigma_{g+w}(x, y) = \inf\{F(z) \mid z \in \mathbb{R}^{n+m+1}, H(z) + (x - \bar{x}) = 0, G(z) + w \leq 0\} \tag{3.42}$$

But, since $H(z) + (x - \bar{x}) = 0$ together with $G(z) + w \leq 0$ is equivalent to the fact that $u = x - \bar{x}$ and $g_j(x, y) + w_j \leq a$, for all $j \in \{0, \ldots, s\}$, it follows that (3.42) can equivalently be written as

$$\inf_{y \in \mathbb{R}^m} \sigma_{g+w}(x, y) = \inf\{a \mid (y, a) \in \mathbb{R}^{m+1}, \sigma_{g+w}(x, y) \leq a\}.$$

The last equality holds, and hence (3.41) is valid.

From now on, let z, F, H, G, ψ be defined as in (3.41). In what follows, we need the notion of upper-semicontinuity. Recall that a set-valued mapping $\mathcal{M}$

from a topological space T into a family of all subsets of $\mathbb{R}^n$ is said to be upper-semicontinuous at $\bar{v} \in T$ if, for any open set $\mathcal{U} \subset \mathbb{R}^n$ with $\mathcal{M}(\bar{v}) \subset \mathcal{U}$, there exists an open neighborhood $\mathcal{V}_{\bar{v}} \subset T$ of $\bar{v}$ such that $\mathcal{M}(v) \subset \mathcal{U}$ whenever $v \in \mathcal{V}_{\bar{v}}$.

Lemma 16. *Condition CC implies Hypothesis H.*

Proof. We prove first that the mapping $(x, w) \rightrightarrows M_{g+w}(x)$ is upper-semicontinuous at all $(\bar{x}, 0) \in \mathbb{R}^n \times \mathbb{R}^{s+1}$. If not, there exists an open set $O \subset \mathbb{R}^m$ with $M(\bar{x}) \subset O$ such that for all open neighborhoods $U_{(\bar{x},0)} \subset \mathbb{R}^n \times \mathbb{R}^{s+1}$ of $(\bar{x}, 0)$ there exists $(x, w) \in U_{(\bar{x},0)}$ with

$$M_{g+w}(x) \not\subset O. \tag{3.43}$$

Now, (3.43) directly implies the existence of sequences $(x_k, w_k)_{k \in \mathbb{N}} \subset \mathbb{R}^n \times \mathbb{R}^{s+1}$ and $(y_k)_{k \in \mathbb{N}} \subset \mathbb{R}^m$ with

$$\begin{cases} (x_k, w_k) \to (\bar{x}, 0), \, k \to \infty, \\ y_k \in M_{g+w_k}(x_k), \text{ for all } k \in \mathbb{N}, \\ y_k \notin O, \text{ for all } k \in \mathbb{N}. \end{cases} \tag{3.44}$$

For $k \in \mathbb{N}$, we have

$$\begin{aligned} |\sigma(\bar{x}, y_k) - \varphi(\bar{x})| &\leq |\sigma(\bar{x}, y_k) - \sigma_{g+w_k}(x_k, y_k)| \\ &\quad + |\sigma_{g+w_k}(x_k, y_k) - \varphi_{g+w_k}(x_k)| + |\varphi_{g+w_k}(x_k) - \varphi(\bar{x})|. \end{aligned} \tag{3.45}$$

Here, the second term on the right-hand side is zero since by (3.44) we have $y_k \in M_{g+w_k}(x_k)$. The first and last terms converge to zero for $k \to \infty$ since $(x_k, w_k) \to (\bar{x}, 0), \, k \to \infty$ (note that $x \mapsto \varphi(x)$ is continuous by condition CC). This implies that $(y_k)_{k \in \mathbb{N}}$ is a minimizing sequence for $\sigma(\bar{x}, \cdot)$,

$$\sigma(\bar{x}, y_k) \to \varphi(\bar{x}), \, k \to \infty.$$

From Lemma 15, condition CC implies condition C*. Hence, by (C2), w.l.o.g., $y_k \to \bar{y} \in M(\bar{x}), \, k \to \infty$. Since $y_k \notin O$ for all $k \in \mathbb{N}$, we see that $\bar{y} \notin O$ (recall that O is open). This is a contradiction to $M(\bar{x}) \subset O$. We conclude that $(x, w) \rightrightarrows M_{g+w}(x)$ is upper-semicontinuous at $(\bar{x}, 0)$.

Now, choose an arbitrary open and bounded set $O \subset \mathbb{R}^m$ with $M(\bar{x}) \subset O$ (note that by (C2) the set $M(\bar{x})$ is bounded). Then, the upper-semicontinuity of $(x, w) \rightrightarrows M_{g+w}(x)$ at $(\bar{x}, 0)$ gives us a neighborhood $U^1_{(\bar{x},0)} \subseteq \mathbb{R}^n \times \mathbb{R}^{s+1}$ of $(\bar{x}, 0)$ such that

$$M_{g+w}(x) \subset O \quad \text{for all } (x, w) \in U^1_{(\bar{x},0)}.$$

This means that for small perturbations all minimizers stay in the compact closure $\bar{O}$.

For the existence of a minimizer, note that due to the openness of condition CC (see Lemma 26) we find an open neighborhood $U^2_{(\bar{x},0)} \subseteq \mathbb{R}^n \times \mathbb{R}^{s+1}$ of $(\bar{x}, 0)$ such that condition CC holds for all $(x, w) \in U^2_{(\bar{x},0)}$ (we can easily modify the proof of

Lemma 26 to see that the openness property also holds for this special class of perturbations—note that we only use the uniform estimate (3.37)).

Now, we can define $\varepsilon > 0$ such that $B_\varepsilon((\bar{x},0)) \subseteq U^1_{(\bar{x},0)} \cap U^2_{(\bar{x},0)}$. Finally, we obtain that for all (x,w) with $\|(x-\bar{x},w)\| < \varepsilon$ the value $\varphi_{g+w}(x)$ is finite and is attained by $\sigma_{g+w}(x,\cdot)$ in O; that is $\psi(x-\bar{x},w)$ is finite and the corresponding NLP has a solution $(x-\bar{x},y,\varphi_{g+w}(x)) \in K$ with

$$K := \overline{B_\varepsilon(\bar{x})} \times \overline{O} \times \bigcup_{(x,w)\in B_\varepsilon((\bar{x},0))} \varphi_{g+w}(x).$$

Due to the continuity of $(x,w) \rightrightarrows \varphi_{g+w}(x)$, the set K is compact. This implies Hypothesis H.$\square$

Recall the notation

$$\sigma(x,y) := \max_{0 \le k \le s} g_k(x,y).$$

Moreover, for $x \in \mathbb{R}^n$, let

$$M(x) := \{y \in \mathbb{R}^m \mid \sigma(x,y) = \varphi(x)\},$$

and, for $(x,y) \in \mathbb{R}^n \times \mathbb{R}^m$, let

$$K(x,y) := \{k \in \{0,\ldots,s\} \mid g_k(x,y) = \sigma(x,y)\}.$$

At a given point $\bar{x} \in \mathbb{R}^n$, we set

$$V(\bar{x}) := \bigcup_{y \in M(\bar{x})} V(\bar{x},y) \subseteq \mathbb{R}^n,$$

where

$$V(\bar{x},y) := \left\{ \sum_{k \in K(\bar{x},y)} \mu_k D_x g_k(\bar{x},y) \;\middle|\; \begin{array}{l} \sum_{k \in K(\bar{x},y)} \mu_k D_y g_k(\bar{x},y) = 0, \\ \sum_{k \in K(\bar{x},y)} \mu_k = 1, \\ \mu_k \ge 0 \end{array} \right\}. \tag{3.46}$$

Lemma 17 (Clarke's subdifferential of φ). *Let condition CC be fulfilled and let $\bar{x} \in \mathbb{R}^n$. Then it holds that*

$$\partial \varphi(\bar{x}) \subseteq conv(V(\bar{x})). \tag{3.47}$$

Proof. Since condition CC holds, we can apply Lemma 16. Using Theorem 27 and (3.41) we obtain the formula for $\partial \varphi(\bar{x})$,

$$\partial \varphi(\bar{x}) = \partial_v \psi(0,0) = \Pi \partial \psi(0,0) \subseteq \Pi \overline{conv(\Delta^1(M_0) + \Delta^0(M_0))}. \tag{3.48}$$

Here, $\Pi : \mathbb{R}^n \times \mathbb{R}^{s+1} \to \mathbb{R}^n$, $(x,w) \mapsto x$, denotes the projection on the first n variables. Note that the identity $\partial_v \psi(0,0) = \Pi \partial \psi(0,0)$ holds due to Clarke's chain rule (see Theorem 2.3.10 in [13]).

We calculate the right-hand side of (3.48);

$$D_z L(z, \delta, \lambda, \mu) = \delta \begin{pmatrix} 0_n \\ 0_{s+1} \\ 1 \end{pmatrix} - \sum_{i=1}^{n} \lambda_i e_i + \sum_{j=1}^{s+1} \mu_j \begin{pmatrix} D_x g_j(u+\bar{x}, y) \\ D_y g_j(u+\bar{x}, y) \\ -1 \end{pmatrix}.$$

If $z = (u, y, a) \in M_0$, we get from $H(z) = -u = 0$ that $u = 0$ and therefore

$$\Delta^\delta(z) = \left\{ (\lambda, \mu) \in \mathbb{R}^n \times \mathbb{R}^{s+1} \;\middle|\; \begin{array}{c} \sum\limits_{j \in K_0(z)} \mu_j = \delta \\ \sum\limits_{j \in K_0(z)} \mu_j D_y g_j(\bar{x}, y) = 0 \\ \sum\limits_{j \in K_0(z)} \mu_j D_x g_j(\bar{x}, y) = \lambda \\ \mu_j \geq 0 \end{array} \right\}.$$

Furthermore, $z \in M_0$ implies that $K_0(z) = K(\bar{x}, y)$, and hence

$$\Pi \Delta^1(z) = \left\{ \sum\limits_{j \in K(\bar{x}, y)} \mu_j D_x g_j(\bar{x}, y) \;\middle|\; \begin{array}{c} \sum\limits_{j \in K(\bar{x}, y)} \mu_j D_y g_j(\bar{x}, y) = 0, \\ \sum\limits_{j \in K(\bar{x}, y)} \mu_j = 1, \\ \mu_j \geq 0 \end{array} \right\}.$$

Note that by the definition of $V(x, y)$ (see (3.46)) and $\Delta^\delta(z)$ we have

$$\Pi \Delta^1(z) = V(\bar{x}, y) \text{ and (trivially) } \Pi \Delta^0(z) = \{0\}.$$

Now, it holds that

$$\Pi \text{conv}(\Delta^1(M_0) + \Delta^0(M_0)) = \text{conv}(\Pi \Delta^1(M_0)) = \text{conv}\left(\bigcup_{z \in M_0} \Pi \Delta^1(z) \right).$$

Together with the fact that $\Delta^1(M_0)$ is bounded, we have

$$\Pi \overline{\text{conv}(\Delta^1(M_0))} = \overline{\Pi \text{conv}(\Delta^1(M_0))}.$$

We obtain

$$\Pi \overline{\text{conv}(\Delta^1(M_0) + \Delta^0(M_0))} = \overline{\text{conv}\left(\bigcup_{z \in M_0} \Pi \Delta^1(z) \right)}$$

$$= \overline{\text{conv}\left(\bigcup_{y \in M(\bar{x})} V(\bar{x}, y) \right)}$$

$$= \overline{\text{conv}(V(\bar{x}))}.$$

Since condition CC holds, we know that $M(\bar{x})$ is compact. This fact implies that $V(\bar{x})$ is compact. Finally, recalling the inclusion in (3.48), we get the assertion.□

Remark 19. It is not known whether the inclusion (3.47) is in fact an equality. This is a topic of future research.

Theorem 28 ($M^{\max}$ is a Lipschitz manifold with boundary). *Let condition CC and the Sym-MFCQ at all points $\bar{x} \in M^{max}$ be fulfilled. Then M^{max} is a Lipschitz manifold (with boundary) of dimension n.*

Proof. Since by Lemma 15 condition C^* holds, we get from Lemma 14 that φ is continuous on $\mathbb{R}^n$. Hence, if $\bar{x} \in M^{\max} = \{x \in \mathbb{R}^n \mid \varphi(x) \geq 0\}$ with $\varphi(\bar{x}) > 0$, then there exists an open neighborhood $U \subseteq \mathbb{R}^n$ of $\bar{x}$ such that $\varphi(x) > 0$ for all $x \in U$. Setting $H(x) := x - \bar{x}$ and $V := H(U)$, we obtain

$$H(\bar{x}) = 0 \text{ and } H(M^{\max} \cap U) = H(U) = \mathbb{R}^n \cap V.$$

Now, let $\bar{x} \in M^{\max} = \{x \in \mathbb{R}^n \mid \varphi(x) \geq 0\}$ with $\varphi(\bar{x}) = 0$. Since the Sym-MFCQ holds, there exists $\xi \in \mathbb{R}^n$ with $v \cdot \xi > 0$ for all $v \in V(\bar{x})$. By applying Lemma 17, we conclude that

$$v \cdot \xi > 0 \text{ for all } v \in \partial\varphi(\bar{x}).$$

Without loss of generality, we assume that $\xi = e_1$ and $\bar{x} = 0$ (otherwise we introduce new coordinates $y := P^{-1} \cdot (x - \bar{x})$, with $P \in R^{n \times n}$ being a rotation matrix with $P \cdot e_1 = \xi$; using Clarke's chain rule (see Theorem 2.3.10 in [13]), we obtain the desired properties for φ in new coordinates). So now we have

$$v \cdot e_1 > 0 \quad \text{for all } v \in \partial\varphi(0).$$

We set H as

$$H(x) := \begin{pmatrix} \varphi(x) \\ x_2 \\ \vdots \\ x_n \end{pmatrix}.$$

Then the generalized Jacobian (according to Clarke; see [13]) at $\bar{x} = 0$ is given by

$$\partial H(0) = \left(\begin{array}{c|ccc} \partial_{x_1}\varphi(0) & \partial_{x_2}\varphi(0) & \dots & \partial_{x_n}\varphi(0) \\ \hline 0 & 1 & & \\ \vdots & & \ddots & \\ 0 & & & 1 \end{array}\right).$$

From Clarke's inverse function theorem (see Theorem 7.1.1 in [13]), we get that there exists an open neighborhood $U \subseteq \mathbb{R}^n$ of $\bar{x}$ such that the inverse H^{-1} exists on U. Moreover, we have with $V := H(U)$

$$H(\bar{x}) = 0 \quad \text{and} \quad H(M^{\max} \cap U) = (\mathbb{H} \times \mathbb{R}^{n-1}) \cap V.$$

This finishes the proof.□

Corollary 8 ($\partial M^{\max}$ is a Lipschitz manifold). *Let condition CC and the Sym-MFCQ be fulfilled. Then, ∂M^{max} is a Lipschitz manifold of dimension $n-1$. Moreover, it holds that*

$$\partial M^{max} = \{x \in \mathbb{R}^n \mid \varphi(x) = 0\}.$$

Proof. The assertion follows directly from the proof of Theorem 28.□

Application to GSIP

We apply our results on the topological properties of $M^{\max}$ in the context of GSIP. It turns out that Assumption B might be replaced by condition CC to obtain the same results as in Section 3.2.1.

Theorem 29 (Sym-MFCQ is generic and stable under CC).
Let $\mathscr{F}$ denote the subset of $\mathscr{C}$ (the set of functions that fulfill condition CC) consisting of those defining functions $(g_0, \dots, g_s)$ for which the Sym-MFCQ holds. Then, $\mathscr{F}$ is C_s^1-open and C_s^1-dense in $\mathscr{C}$.

Proof. The proof runs along the same lines as the proof of Theorem 22. condition CC implies a local description of φ as it is shown to hold in Lemma 7. Furthermore, the set-valued mappings $(x, g) \rightrightarrows M_g(x)$ and $(x, y, g) \rightrightarrows V_g(x, y)$ can be proven to be upper-semicontinuous with the same arguments as used in the proof of Lemma 8. This, together with the compactness of $M_g(x)$ for $(x, g) \in \mathbb{R}^n \times \mathscr{C}$, implies that the Sym-MFCQ is locally stable. The globalization procedure is standard.□

Theorem 30 (Closure theorem under CC). *Let condition CC and the Sym-MFCQ at all points $\bar{x} \in M^{max}$ hold. Then,*

$$\overline{M} = M^{max}.$$

Proof. See the proof of Theorem 23. Again, the main property that is used there is the local description of $M^{\max}$. As we have seen above, this description is also valid under condition CC.□

As a direct consequence of Theorems 28 and 30, we describe the topological structure of $\overline{M}$ in a generic situation.

Theorem 31 (Closure of the GSIP feasible set is Lipschitz). *Let condition CC and the Sym-MFCQ at all points $\bar{x} \in M^{max}$ be fulfilled. Then, the closure of the GSIP feasible set $\overline{M}$ is a Lipschitz manifold (with boundary) of dimension n.*

The main reason to introduce condition CC instead of Assumption B is the result in Theorem 31. Moreover, Assumption B is not symmetric w.r.t. defining functions $g_0, \dots, g_s$. It does not involve the function g_0. This issue may cause some undesirable effects, as the following example shows.

Example 23 (Assumption B does not imply Hypothesis H). Let $m = n = 1$ and $s = 1$. Consider a function $g_1 \in C^1(\mathbb{R} \times \mathbb{R})$ with $g_1(x,y) \leq C, C > 0$, that fulfills Assumption B (i.e., the set-valued mapping $x \rightrightarrows \{y \in \mathbb{R} \mid g_1(x,y) \leq 0\}$ is locally bounded). Then, let $g_0 \in C^1(\mathbb{R} \times \mathbb{R})$ be a function with the following properties:

$$g_0(x,y) > \max\{g_1(x,y), C, 0\} \text{ for all } (x,y) \in \mathbb{R} \times \mathbb{R}$$
$$g_0(x,y) \to C, |y| \to \infty \text{ for all } x \in \mathbb{R}.$$

Now, we have

$$\varphi(x) = \inf_{y \in \mathbb{R}} \sigma(x,y) = \inf_{y \in \mathbb{R}} g_0(x,y) = C,$$

and since the infimum is not attained, Hypothesis H does not hold.

We conclude that condition CC is a natural symmetric assumption for the related optimization problem on the closure of the GSIP feasible set

$$\overline{\text{GSIP}} : \quad \text{minimize } f(x) \text{ s.t. } x \in \overline{M}.$$

Nonsmooth analysis perspective

Lemma 17 and Theorem 29 can be interpreted in terms of nonsmooth analysis. In fact, under CC we get

$$\partial \varphi(\bar{x}) \subseteq \text{conv}(V(\bar{x})).$$

A generic and stable Sym-MFCQ provides the existence of a vector $\xi \in \mathbb{R}^n$ such that

$$V(\bar{x}) \cdot \xi > 0.$$

From this, there exists generically a vector $\xi \in \mathbb{R}^n$ such that

$$\partial \varphi(\bar{x}) \cdot \xi > 0.$$

This means that Clarke's subdifferential of a min-max function is generically regular. Moreover, the zero-set of a min-max function defined on $\mathbb{R}^n$ is Lipschitz homeomorphic to $\mathbb{R}^{n-1}$. This is a considerable generalization of corresponding results from transversality theory and smooth analysis (see, e.g., [63]), where it is for instance commonly known that under generic assumptions the zero-set of a smooth function defined on $\mathbb{R}^n$ is locally diffeomorphic to $\mathbb{R}^{n-1}$. This motivates a further investigation of other types of nonsmoothness in order to derive similar results.

3.3 Nonsmooth symmetric reduction ansatz

As we have seen in Section 3.2.1, the feasible set M in GSIPs need not be closed. Under the Sym-MFCQ its closure $\overline{M}$ can be described by means of infinitely many

inequality constraints of maximum type. In this section, we introduce the nonsmooth symmetric reduction ansatz (NSRA). Under the NSRA, we prove that the set
$\overline{M}$ can be described locally as the feasible set of a disjunctive optimization problem
defined by finitely many inequality constraints of maximum type. This also shows
the appearance of re-entrant corners in $\overline{M}$. Under the Sym-MFCQ, all local minimizers of GSIP are KKT points (for GSIPs). We show that the NSRA is generic and
stable at all KKT points and that all KKT points are nondegenerate. The concept of
nondegenerate KKT points as well as a corresponding GSIP-index are introduced.
In particular, a nondegenerate KKT point is a local minimizer if and only if its
GSIP-index vanishes. At local minimizers, the NSRA coincides with the symmetric reduction ansatz (SRA) as introduced in [37]. In comparison with the SRA, the
main new issue in the NSRA is that at KKT points different from local minimizers
the Lagrange polytope at the lower level generically need no longer be a singleton.
In fact, it will be a full dimensional simplex. This fact is crucial to providing the
above mentioned local reduction to a disjunctive optimization problem. We refer
the reader to [73] for details.

Formulation of NSRA

Recall that from Assumption B and the Sym-MFCQ on $M^{\max}$ we obtain

$$\overline{M} = M^{\max}, \tag{3.49}$$

where

$$M^{\max} = \{x \in \mathbb{R}^n \mid \sigma(x,y) \geq 0 \text{ for all } y \in \mathbb{R}^m\}.$$

We consider the relaxed problem

$$\overline{\text{GSIP}}: \quad \text{minimize } f(x) \text{ s.t. } x \in \overline{M}. \tag{3.50}$$

Its feasible set $\overline{M}$ is given by infinitely many constraints of maximum type (cf.
(3.49)).

Our main goal is to provide a reduced local description of $\overline{M}$. To this aim, the
nonsmooth symmetric reduction ansatz (NRSA) will be introduced.

Let $\bar{x} \in \overline{M}$. We set as before

$$M(\bar{x}) := \{y \in \mathbb{R}^m \mid \sigma(\bar{x},y) = 0\}.$$

Note that $M(\bar{x})$ consists of the global minimizers of $\sigma(\bar{x},\cdot)$ with vanishing optimal value. We consider the well-known epigraph reformulation that $\bar{y}$ is a global
minimizer of $\sigma(\bar{x},\cdot)$ with vanishing optimal value if and only if $(\bar{y},0)$ is a global
minimizer of

$$Q(\bar{x}): \quad \min_{(y,z) \in \mathbb{R}^m \times \mathbb{R}} z \quad \text{s.t.} \quad z - g_k(\bar{x},y) \geq 0, \, k = 0, \ldots s.$$

From the first-order optimality condition for $(\bar{y}, 0)$, we obtain that the corresponding polytope of Lagrange multipliers $\Delta(\bar{x}, \bar{y})$ is nonempty:

$$\Delta(\bar{x}, \bar{y}) := \left\{ (\gamma_k)_{k \in K_0(\bar{x}, \bar{y})} \in \mathbb{R}^{|K_0(\bar{x}, \bar{y})|} \left| \begin{array}{l} \sum\limits_{k \in K_0(\bar{x}, \bar{y})} \gamma_k D_y g_k(\bar{x}, \bar{y}) = 0, \\ \sum\limits_{k \in K_0(\bar{x}, \bar{y})} \gamma_k = 1, \ \gamma_k \geq 0, \ k \in K_0(\bar{x}, \bar{y}) \end{array} \right. \right\}.$$

$K_0(\bar{x}, \bar{y}) := \{k \in \{0, \ldots, s\} \mid g_k(\bar{x}, \bar{y}) = 0\}$ is the active index set for $(\bar{y}, 0)$.

For $\gamma \in \Delta(\bar{x}, \bar{y})$, we set

$$K_+(\gamma) := \{k \in K_0(\bar{x}, \bar{y}) \mid \gamma_k > 0\}.$$

For $\bar{x} \in \overline{M}$ and $\bar{y} \in M(\bar{x})$, we define the finite set

$$\mathscr{E}(\bar{x}, \bar{y}) := \{\gamma \mid \gamma \text{ is a vertex of the polytope } \Delta(\bar{x}, \bar{y})\}.$$

Now, we are ready to state the nonsmooth symmetric reduction ansatz.

Definition 26 (NSRA). The nonsmooth symmetric reduction ansatz (NSRA) is said to hold at $\bar{x} \in \overline{M}$ if for every $\bar{y} \in M(\bar{x})$ either Case I or Case II occurs.

Case I: $(\bar{y}, 0)$ is a nondegenerate minimizer for $Q(\bar{x})$.

Case II: $\Delta(\bar{x}, \bar{y})$ is not a singleton and
$|K_+(\gamma)| = m + 1$ for all vertices $\gamma \in \mathscr{E}(\bar{x}, \bar{y})$.

We give three guiding remarks on the Cases I and II.

Remark 20 (Case I in NSRA). Case I means that linear independence constraint qualification (LICQ), strict complementarity slackness (SC) and second-order sufficiency condition (SOSC) hold at the solution $(\bar{y}, 0)$ of $Q(\bar{x})$. This corresponds to the symmetric reduction ansatz (SRA) as introduced in [37]. In Case I we obtain, in particular, that $\Delta(\bar{x}, \bar{y}) = \{\bar{\gamma}\}$ is a singleton (due to the LICQ) and $K_+(\bar{\gamma}) = K_0(\bar{x}, \bar{y})$ (due to the SC).

Remark 21 (Case II in NSRA). In Case II, the polytope $\Delta(\bar{x}, \bar{y})$ is not a singleton due to the possible violation of the LICQ for $Q(\bar{x})$. Nevertheless, for a vertex $\gamma \in \mathscr{E}(\bar{x}, \bar{y})$, we may define the truncated optimization problem

$$Q^\gamma(\bar{x}) : \quad \min_{(y,z) \in \mathbb{R}^m \times \mathbb{R}} z \quad \text{s.t.} \quad z - g_k(\bar{x}, y) \geq 0, \ k \in K_+(\gamma).$$

The crucial fact here is that $(\bar{y}, 0)$ is a nondegenerate local minimizer of $Q^\gamma(\bar{x})$ (see Lemma 18 for details). Indeed, for $Q^\gamma(\bar{x})$, the LICQ follows from the fact that $K_+(\gamma)$ is minimal w.r.t. inclusion among all the sets $K_+(\delta)$, $\delta \in \Delta(\bar{x}, \bar{y})$. Strict complementarity slackness holds from the definition of $K_+(\gamma)$. The "full-dimensionality" condition $|K_+(\gamma)| = m + 1$ implies that the corresponding tangent space shrinks to a point (the origin) and hence the SOSC holds trivially. Moreover, the family of parameterized nondegenerate optimization problems

$$\{Q^{\gamma}(\bar{x}) \mid \gamma \in \mathscr{E}(\bar{x}, \bar{y})\}$$

will lead us to the reduction result on the local description of $\overline{M}$ as a so-called disjunctive optimization problem (see Theorem 32).

Remark 22 (SRA). In [37], the so-called symmetric reduction ansatz (SRA) is introduced. The main difference between the SRA and NSRA is that the SRA only focuses on Case I in the definition of the NSRA. In [38], it is shown that the SRA holds generically at all local minimizers. In order to extend the idea of reduction to all KKT points (see Definition 20 below), Case II in the NSRA is crucial, and its appearance cannot be avoided (see Example 25).

Next, we recall the notion of a Karush-Kuhn-Tucker point for $\overline{GSIP}$ from Definition 20.

The main results concerning the NSRA and its impacts on the local description of $\overline{M}$ are the following:

(i)   Under the NRSA, the set $\overline{M}$ can be locally described as in disjunctive optimization (see Theorem 32). Using the corresponding results on disjunctive optimization Problems (see [71]), we introduce the notions of a nondegenerate KKT point for $\overline{GSIP}$ and its GSIP-index (see Definitions 27 and 28). In particular, the GSIP-index vanishes if and only if the corresponding point is a local minimizer for $\overline{GSIP}$.

(ii) The NSRA is proven to hold generically at all KKT points. Moreover, all KKT points are proven to be generically nondegenerate (see Theorem 33).

(iii) The NSRA is shown to be stable under C^2-perturbations of the defining functions at nondegenerate KKT points (see Theorem 33).

Remark 23. The main result (iii) above shows in particular that the concept of the SRA as introduced in [37] is stable at local minimizers.

Reduction under NSRA

First, we state Theorem 32 on the local reduction of $\overline{M}$ under the NRSA.

Theorem 32 (Local reduction). *Let the Sym-MFCQ hold at all points of M^{max} and Assumption B be satisfied. Let the NSRA hold at $\bar{x} \in \overline{M}$. Then,*

(i)   *$M(\bar{x})$ is finite, w.l.o.g. $M(\bar{x}) = \{\bar{y}_1, \ldots, \bar{y}_p, \bar{y}_{p+1}, \ldots, \bar{y}_l\}$, where for $\bar{y}_1, \ldots, \bar{y}_p$ Case I and for $\bar{y}_{p+1}, \ldots, \bar{y}_l$ Case II from the NSRA occurs, and*

(ii) *there exist an open neighborhood U of $\bar{x}$ and implicit functions $(y_i, z_i) : U \longrightarrow \mathbb{R}^m \times \mathbb{R}$, $i = 1, \ldots, p$ and $(y_j^{\gamma}, z_j^{\gamma}) : U \longrightarrow \mathbb{R}^m \times \mathbb{R}$, $\gamma \in \mathscr{E}(\bar{x}, \bar{y}_j)$, $j = p+1, \ldots, l$, such that*

(a) $(y_i, z_i)(\bar{x}) = (\bar{y}_i, 0)$ and $(y_j^\gamma, z_j^\gamma)(\bar{x}) = (\bar{y}_j, 0)$,

(b) $(y_i(x), z_i(x))$ is the locally unique local minimizer of $Q(x)$,
       $(y_j^\gamma(x), z_j^\gamma(x))$ is the locally unique local minimizer of $Q^\gamma(x)$, and

(c) $y_i(\cdot), y_j^\gamma(\cdot)$ are at least once and $z_i(\cdot), z_j^\gamma(\cdot)$ twice continuously differen-
       tiable.

Moreover, it holds that

$$\overline{M} \cap U = \left\{ x \in U \,\middle|\, \begin{array}{l} z_i(x) \geq 0,\ i = 1, \ldots, p \\ \displaystyle\max_{\gamma \in \mathscr{E}(\bar{x}, \bar{y}_j)} z_j^\gamma(x) \geq 0,\ j = p+1, \ldots, l \end{array} \right\}. \tag{3.51}$$

Remark 24. Note that the "max-inequalities" in (3.51) give rise to the well-known appearance of re-entrant corners in $\overline{M}$ (see [110]). The representation (3.51) of $\overline{M}$ is due to the special structure of the Lagrange polytope $\Delta(\bar{x}, \bar{y})$ in Case II. The observation that the description of the GSIP feasible set is connected with the lower-level multipliers is from [109].

From Theorem 32, we see that under the NSRA at $\bar{x}$ the optimization problem $\overline{\text{GSIP}}$ is locally equivalent to the following reduced problem:

$$\overline{\text{GSIP}}_{\text{red}} : \ \text{minimize}\ f(x)\ \text{s.t.}\ z_i(x) \geq 0,\ i = 1, \ldots, p, \\ \max_{\gamma \in \mathscr{E}(\bar{x}, \bar{y}_j)} z_j^\gamma(x) \geq 0,\ j = p+1, \ldots, l. \tag{3.52}$$

The feasible set of $\overline{\text{GSIP}}_{\text{red}}$ is given by the finite number of inequality constraints and the finite number of maximum-type constraints with twice continuously differentiable data functions. $\overline{\text{GSIP}}_{\text{red}}$ is well-known to be referred to as a disjunctive optimization problem (see [71]). Although its data functions $z_i(\cdot), z_j^\gamma(\cdot)$ are defined implicitly, we can explicitly obtain their first- and second-order derivatives at the point of interest $\bar{x}$.

Remark 25. We recall that $(\bar{y}_i, 0)$ is a nondegenerate minimizer of $Q(\bar{x})$ (see Case I). Moreover, $(\bar{y}_j, 0)$ is a nondegenerate minimizer of $Q^\gamma(\bar{x})$, $\gamma \in \mathscr{E}(\bar{x}, \bar{y}_j)$ (see Lemma 18). Hence, implicit functions $z_i(\cdot), z_j^\gamma(\cdot)$ in Theorem 32 can be obtained from standard results on parametric nonlinear optimization. In fact, $z_i(\cdot)$ and $z_j^\gamma(\cdot)$ are local optimal value functions for $Q(\bar{x})$ and $Q^\gamma(\bar{x})$, respectively. Thus, we can explicitly obtain their first- and second-order derivatives at $\bar{x}$ (see, e.g., [9, 65]):

$$Dz_i(\bar{x}) = \sum_{k \in K_0(\bar{x}, \bar{y}_i)} \bar{\gamma}_k^i D_x g_k(\bar{x}, \bar{y}_i) \ \text{with}\ \Delta(\bar{x}, \bar{y}_i) = \{\bar{\gamma}^i\} \ \text{for}\ i = 1, \ldots, p,$$

$$Dz_j^\gamma(\bar{x}) = \sum_{k \in K_+(\gamma)} \gamma_k D_x g_k(\bar{x}, \bar{y}_j) \quad \text{for}\ \gamma \in \mathscr{E}(\bar{x}, \bar{y}_j),\ j = p+1, \ldots, l.$$

Note that $Dz_i(\bar{x}) \in V(\bar{x}, \bar{y}_i)$ and $Dz_j^\gamma(\bar{x}) \in V(\bar{x}, \bar{y}_j)$.

Setting $K_0 := K_0(\bar{x}, \bar{y}_i)$ and evaluating at $(\bar{x}, \bar{y}_i)$, we get

$$D^2 z_i(\bar{x}) = \sum_{k \in K_0} \bar{\gamma}_k^i D_{xx}^2 g_k - A^T \cdot B^{-1} \cdot A, \text{ where}$$

$$A := \begin{pmatrix} \sum_{k \in K_0} \bar{\gamma}_k^i D_{xy} g_k \\ 0 \\ -D_x g_k, k \in K_0 \end{pmatrix}, B := \begin{pmatrix} \sum_{k \in K_0} \bar{\gamma}_k^i D_{yy}^2 g_k & 0 & -D_y^T g_k, k \in K_0 \\ 0 & 0 & 1 \\ -D_y g_k, k \in K_0 & 1 & 0 \end{pmatrix}.$$

For $D^2 z_j^\gamma(\bar{x})$, a similar formula holds, where K_0 has to be replaced by $K_+(\gamma)$.

The formulas above can be used for explicit formulations of optimality criteria for GSIPs (see [41]).

We consider the notion of a stationary point for the disjunctive optimization problem $\overline{GSIP}_{red}$ as defined in [71]. The point $\bar{x}$ is called stationary for $\overline{GSIP}_{red}$ if there exist $\bar{\lambda}_i \geq 0, i = 1, \ldots, p$ and $\bar{\lambda}_j^\gamma \geq 0, \gamma \in \mathscr{E}(\bar{x}, \bar{y}_j), j = p+1, \ldots, l$, such that

$$Df(\bar{x}) = \sum_{i=1}^p \bar{\lambda}_i Dz_i(\bar{x}) + \sum_{j=p+1}^l \sum_{\gamma \in \mathscr{E}(\bar{x}, \bar{y}_j)} \bar{\lambda}_j^\gamma Dz_j^\gamma(\bar{x}). \tag{3.53}$$

Note that all constraints in (3.52) are active at $\bar{x}$ from Theorem 32 (i).

The following crucial observation is from Remark 25. The point $\bar{x}$ is stationary for $\overline{GSIP}_{red}$ if and only if it is a KKT point for $\overline{GSIP}$ according to Definition 20. This fact gives rise to introducing the notions of a nondegenerate KKT point and its GSIP-index along the lines in [71].

Definition 27 (Nondegenerate KKT point). Let $\bar{x} \in \overline{M}$ be a KKT point for $\overline{GSIP}$ according to Definition 20. Then, $\bar{x}$ is called nondegenerate if the following conditions are satisfied:

ND1:   NSRA holds at $\bar{x}$.
ND2:   The vectors

$$Dz_i(\bar{x}), i = 1, \ldots, p, Dz_j^\gamma(\bar{x}), \gamma \in \mathscr{E}(\bar{x}, \bar{y}_j), j = p+1, \ldots, l$$

are linearly independent (i.e. LICQ for $\overline{GSIP}_{red}$).
ND3:   The uniquely determined (due to ND2) multipliers in (3.53)

$$\bar{\lambda}_i, i = 1, \ldots, p, \bar{\lambda}_j^\gamma, \gamma \in \mathscr{E}(\bar{x}, \bar{y}_j), j = p+1, \ldots, l$$

are positive (i.e., SC for $\overline{GSIP}_{red}$).
ND4:   The matrix
$$V^T \cdot D^2 L(\bar{x}) \cdot V$$

is nonsingular, where $D^2 L(\bar{x})$ stands for the Hessian of the Lagrange function

$$L(x) = f(x) - \sum_{i=1}^p \bar{\lambda}_i z_i(x) + \sum_{j=p+1}^l \sum_{\gamma \in \mathscr{E}(\bar{x}, \bar{y}_j)} \bar{\lambda}_j^\gamma z_j^\gamma(x) \tag{3.54}$$

and V is a matrix whose columns form a basis of the tangent space

$$\{\xi \in \mathbb{R}^n \mid Dz_i(\bar{x}) \cdot \xi = 0, i = 1, \ldots, p,$$
$$Dz_j^\gamma(\bar{x}) \cdot \xi = 0, \gamma \in \mathscr{E}(\bar{x}, \bar{y}_j), j = p+1, \ldots, l\}. \tag{3.55}$$

Definition 28 (GSIP-index [71, Definition 2.3]). Let $\bar{x} \in \overline{M}$ be a nondegenerate KKT point for $\overline{\text{GSIP}}$. We call the number of negative eigenvalues of the matrix $V^T \cdot D^2 L(\bar{x}) \cdot V$ from ND4 the quadratic index of $\bar{x}$ and denote it by QI. Furthermore, we call the number

$$\text{QI} + \sum_{j=p+1}^{l} \left[\left| \mathscr{E}(\bar{x}, \bar{y}_j) \right| - 1 \right]$$

the GSIP-index of $\bar{x}$.

In particular, the point $\bar{x} \in \overline{M}$ is a local minimizer for the GSIP if and only if its GSIP-index vanishes (see [71]).

Let $\mathscr{A}$ denote the set of problem data $(f, g_0, \ldots, g_s) \in C^2(\mathbb{R}^n) \times \left[C^2(\mathbb{R}^n \times \mathbb{R}^m)\right]^{s+1}$ such that Assumption B is satisfied. The set $\mathscr{A}$ is C_s^0-open (see [68]).

Let $\mathscr{B}$ denote the set of problem data $(f, g_0, \ldots, g_s) \in C^2(\mathbb{R}^n) \times \left[C^2(\mathbb{R}^n \times \mathbb{R}^m)\right]^{s+1}$ such that the Sym-MFCQ is satisfied at all points of $M^{\max}$. The set $\mathscr{B}$ is C_s^1-open and C_s^1-dense in $\mathscr{A}$ (see Theorem 22).

Theorem 33 (Nondegenerate KKT points are generic and stable). *Let $\mathscr{F}$ denote the subset of $\mathscr{A} \cap \mathscr{B}$ consisting of those problem data $(f, g_0, \ldots, g_s)$ for which all KKT points are nondegenerate. Then, $\mathscr{F}$ is C_s^2-open and C_s^2-dense in $\mathscr{A}$.*

Recall that the NSRA holds at a nondegenerate KKT point according to Definition 27. Hence, in particular, the NSRA is a generic and stable condition at KKT points.

As an illustration, we give two examples of Cases I and II in the NSRA. First, we provide Example 24, for which the NSRA does not hold. By special perturbations of the defining functions, one can attain Case I. Here, for the sake of explanation, we restrict our considerations to the particular Case of the SIP (see also the discussion below). Second, in Example 25 (from [40, 110]), Case II turns out to be stable under arbitrary small C^1-perturbations of defining functions.

Example 24 (Density of Case I in NSRA). Let $n = 0$, $m = 2$, $s = 2$, and the GSIP be given by

$$g_0(y_1, y_2) = -y_1 + y_2^2, \, g_1(y_1, y_2) = y_1 + y_2^2, \, g_2(y_1, y_2) = 2y_1 + y_2^2.$$

We consider the global minimizer $(0,0)$ of $\sigma(y_1, y_2) = \max_{k \in \{0,1,2\}} g_k(y_1, y_2)$. The vectors $(-Dg_k(0,0), 1)^T$, $k \in \{0,1,2\}$ are linearly dependent. Hence, Case I in the NSRA does not hold at $(0,0)$. The vertices of $\Delta(0,0)$ are $\gamma_1 = \left(\frac{1}{2}, \frac{1}{2}, 0\right)$ and $\gamma_2 = \left(\frac{2}{3}, 0, \frac{1}{3}\right)$. Thus, $|K_+(\gamma_1)| = |K_+(\gamma_2)| = 2 \neq 3 = m+1$, and Case II in the NSRA does not hold at $(0,0)$.

For sufficiently small $\varepsilon > 0$, we perturb the functions g_0, g_1, g_2 as

$$g_0^\varepsilon(y_1, y_2) = -y_1 + y_2^2 - \varepsilon y_2, \; g_1^\varepsilon(y_1, y_2) = y_1 + y_2^2 + \varepsilon y_2, \; g_2^\varepsilon(y_1, y_2) = 2y_1 + y_2^2 + 3\varepsilon y_2.$$

We consider GSIP^ε with defining functions $g_0^\varepsilon, g_1^\varepsilon, g_2^\varepsilon$. The vectors $\left(-Dg_k^\varepsilon(0,0), 1 \right)^T$, $k \in \{0, 1, 2\}$ now become linearly independent. Moreover, $\Delta^\varepsilon(0,0) = \left\{ \left(\frac{5}{8}, \frac{1}{8}, \frac{2}{8} \right) \right\}$. This implies that $(0,0)$ is the nondegenerate global minimizer of $\sigma^\varepsilon(y_1, y_2) = \max\limits_{k \in \{0,1,2\}} g_k^\varepsilon(y_1, y_2)$. Hence, Case I in become NSRA occurs at $(0,0)$ for GSIP^ε.

Example 25 (Stability of Case II in NSRA). Let $n = 2$, $m = 1$, $s = 2$, and the GSIP be given by

$$g_0(x_1, x_2, y) = y, \; g_1(x_1, x_2, y) = x_1 - y, \; g_2(x_1, x_2, y) = x_2 - y.$$

We consider for $(\bar{x}_1, \bar{x}_2) = (0,0)$ the global minimizer 0 of the function $\sigma(\bar{x}_1, \bar{x}_2, y) = \max\limits_{k \in \{0,1,2\}} g_k(\bar{x}_1, \bar{x}_2, y)$. The vertices of $\Delta(0,0,0)$ are $\gamma_1 = \left(\frac{1}{2}, 0, \frac{1}{2} \right)$ and $\gamma_2 = \left(\frac{1}{2}, \frac{1}{2}, 0 \right)$. Thus, $|K_+(\gamma_1)| = |K_+(\gamma_2)| = 2 = m + 1$, and Case II in the NSRA holds at 0. Note that the feasible set M of the GSIP is given by

$$M = \left\{ (x_1, x_2) \in \mathbb{R}^2 \,|\, \max\{x_1, x_2\} \geq 0 \right\}$$

and possesses the disjunctive structure. We point out that the validity of Case II here is stable under arbitrary small C^1-perturbations of the defining functions g_0, g_1, g_2.

Proofs of main results

Lemmas 18, 19 and 20 deal with the parametric problems $Q(\cdot)$ and $Q^\gamma(\cdot)$ corresponding to Case II in the NSRA.

Lemma 18 (Nondegeneracy for $Q^\gamma(\bar{x})$ in Case II). *Let the NSRA hold at $\bar{x} \in \overline{M}$, and for $\bar{y} \in M(\bar{x})$ let Case II occur. Given a vertex $\gamma \in \mathcal{E}(\bar{x}, \bar{y})$, $(\bar{y}, 0)$ is a nondegenerate local minimizer of the truncated optimization problem*

$$Q^\gamma(\bar{x}): \quad \min_{(y,z) \in \mathbb{R}^m \times \mathbb{R}} z \quad s.t. \quad z - g_k(\bar{x}, y) \geq 0, \; k \in K_+(\gamma).$$

Proof. Note that $(\bar{y}, 0)$ is a KKT point for $Q^\gamma(\bar{x})$ with the Lagrange multiplier vector γ. We show that the LICQ, SC, and SOSC are fulfilled at $(\bar{y}, 0)$.

(a) Assume that the LICQ does not hold at $(\bar{y}, 0)$. Then, there exist real numbers β_k, $k \in K_+(\gamma)$ (not all vanishing) such that

$$\sum_{k \in K_+(\gamma)} \beta_k D_y g_k(\bar{x}, \bar{y}) = 0, \quad \sum_{k \in K_+(\gamma)} \beta_k = 0. \tag{3.56}$$

We claim that there exists a real number $a \neq 0$ such that

$$\gamma_k + a\beta_k \geq 0 \text{ for all } k \in K_+(\gamma) \text{ and}$$

$$\gamma_i + a\beta_i = 0 \text{ for at least one } i \in K_+(\gamma). \tag{3.57}$$

Indeed, put $\tau := \min\left\{ \frac{\gamma_k}{|\beta_k|} \mid \beta_k \neq 0 \right\} > 0$ and let the minimum be attained at some index l. We define

$$a := \begin{cases} \tau & \text{if } \beta_l < 0, \\ -\tau & \text{if } \beta_l > 0. \end{cases}$$

From (3.56) and $\gamma \in \Delta(\bar{x}, \bar{y})$, we obtain

$$\sum_{k \in K_+(\gamma)} (\gamma_k + a\beta_k) D_y g_k(\bar{x}, \bar{y}) = 0, \quad \sum_{k \in K_+(\gamma)} \gamma_k + a\beta_k = 1.$$

Hence, $\gamma_k + a\beta_k \in \Delta(\bar{x}, \bar{y})$. Moreover, $K_+(\gamma_k + a\beta_k)$ is a proper subset of $K_+(\gamma)$ from (3.57). However, since $\gamma \in \mathscr{E}(\bar{x}, \bar{y})$ is a vertex of $\Delta(\bar{x}, \bar{y})$, $K_+(\gamma)$ is minimal w.r.t. inclusion among all the sets $K_+(\delta)$, $\delta \in \Delta(\bar{x}, \bar{y})$. This fact provides a contradiction.

(b) Strict complementarity slackness holds from the definition of $K_+(\gamma)$.

(c) Since $|K_+(\gamma)| = m + 1$ and all constraints $z - g_k(\bar{x}, y)$, $k \in K_+(\gamma)$ are active at $(\bar{y}, 0)$, the corresponding tangent space for $Q^\gamma(\bar{x})$ vanishes. Hence, the SOSC holds trivially. $\square$

For the following, we need the concept of a strongly stable (in the sense of Kojima [83]) KKT point for nonlinear programming. We refer the reader to Section 1.4 for the definition of a strongly stable KKT point and its characterization.

Lemma 19 (Strong Stability for $Q(\bar{x})$ in Case II). *Let the NSRA hold at $\bar{x} \in \overline{M}$, and for $\bar{y} \in M(\bar{x})$ let Case II occur. Then, $(\bar{y}, 0)$ is a strongly stable (local) minimizer of $Q(\bar{x})$.*

Proof. We use the characterization of a strongly stable KKT point given in Theorem 9 (see [34]). First, we note that the standard Mangasarian-Fromovitz constraint qualification (MFCQ) is fulfilled at $(\bar{y}, 0)$ by the epigraph reformulation $Q(\bar{x})$. Since $\Delta(\bar{x}, \bar{y})$ is not a singleton in Case II, the LICQ fails to hold at $(\bar{y}, 0)$. Thus, $(\bar{y}, 0)$ is a strongly stable minimizer of $Q(\bar{x})$ if and only if

$$D^2_{(y,z)} L(\bar{y}, 0, \delta) \text{ is positive definite on } T^{K_+(\delta)} \text{ for all } \delta \in \Delta(\bar{x}, \bar{y}), \tag{3.58}$$

where $D^2_{(y,z)} L(\bar{y}, 0, \delta)$ stands for the Hessian of the Lagrange function

$$L(y, z, \delta) = z - \sum_{k \in K_0(\bar{x}, \bar{y})} \delta_k (z - g_k(\bar{x}, y)) = \sum_{k \in K_0(\bar{x}, \bar{y})} \delta_k g_k(\bar{x}, y)$$

and $T^{K_+(\delta)}$ is the subspace

$$\left\{ \begin{pmatrix} \xi \\ \eta \end{pmatrix} \in \mathbb{R}^m \times \mathbb{R} \mid \left(-D_y g_k(\bar{x}, \bar{y}) \; 1 \right) \cdot \begin{pmatrix} \xi \\ \eta \end{pmatrix} = 0, k \in K_+(\delta) \right\}.$$

For a vertex $\gamma \in \mathscr{E}(\bar{x}, \bar{y})$, the vectors

$$\begin{pmatrix} -D_y^T g_k(\bar{x}, \bar{y}) \\ 1 \end{pmatrix}, \ k \in K_+(\gamma),$$

are linearly independent (see part (a) in the proof of Lemma 18). Hence, $T^{K_+(\gamma)} = \{0\}$ since $|K_+(\gamma)| = m + 1$. Moreover, $K_+(\gamma)$ is minimal w.r.t. inclusion among all the sets $K_+(\delta)$, $\delta \in \Delta(\bar{x}, \bar{y})$. Thus, we have that $T^{K_+(\delta)} \subset T^{K_+(\gamma)} = \{0\}$ for all $\delta \in \Delta(\bar{x}, \bar{y})$, and (3.58) is trivially satisfied. $\square$

Lemma 20 (Active index set for $Q(\cdot)$ in Case II). *Let the NSRA hold at $\bar{x} \in \overline{M}$, and for $\bar{y} \in M(\bar{x})$ let Case II occur. Let $(y(x), z(x))$ be the locally unique local minimizer of $Q(x)$ (existing due to Lemma 19) with the active index set*

$$K_0(x) := \{k \in \{0, \ldots, s\} \,|\, z(x) - g_k(x, y(x)) = 0\}.$$

Then, for all x sufficiently close to $\bar{x}$, it holds that $K_0(x) \in \mathcal{K}(\bar{x}, \bar{y})$, where

$$\mathcal{K}(\bar{x}, \bar{y}) := \{K \subset K_0(\bar{x}, \bar{y}) \,|\, \text{ there exists } \gamma \in \Delta(\bar{x}, \bar{y}) \text{ with } K_+(\gamma) \subset K\}.$$

Proof. Note that the function $(y(\cdot), z(\cdot))$ is locally Lipschitz continuous due to the strong stability of $(\bar{y}, 0)$ for $Q(\bar{x})$. Moreover, $K_0(\bar{x}) = K_0(\bar{x}, \bar{y})$. Hence, for x sufficiently close to $\bar{x}$, it holds that $K_0(x) \subset K_0(\bar{x}, \bar{y})$. In particular, for x there exists a sequence $(x_i)_{i \in \mathbb{N}}$ such that

$$x_i \xrightarrow{i} \bar{x} \text{ and } K_0(x_i) = K_0(x).$$

Let $\gamma(x_i)$ be a vector of Lagrange multipliers for $(y(x_i), z(x_i))$,

$$\sum_{k \in K_0(x)} \gamma_k(x_i) D_y g_k(x_i, y(x_i)) = 0, \ \sum_{k \in K_0(x)} \gamma_k(x_i) = 1, \ \gamma_k(x_i) \geq 0, \ k \in K_0(x). \quad (3.59)$$

We set $\gamma_k(x_i) = 0$ for $k \notin K_0(x)$. The corresponding polytopes of Lagrange multipliers $\Delta(x_i, y(x_i))$ (consisting of those $\gamma(x_i)$ in (3.59)) are compact. Moreover, since their vertices are depending continuously on x_i and $x_i \xrightarrow{i} \bar{x}$, there exists a compact set $V \subset \mathbb{R}^{|K_0(\bar{x}, \bar{y})|}$ such that

$$\bigcup_{i \in \mathbb{N}} \Delta(x_i, y(x_i)) \subset V.$$

Without loss of generality, we may assume that $\gamma(x_i) \xrightarrow{i} \bar{\gamma} \in V$. Obviously, $K_+(\bar{\gamma}) \subset K_0(x)$. Letting $i \longrightarrow \infty$ in (3.59), we obtain that $\bar{\gamma} \in \Delta(\bar{x}, \bar{y})$. $\square$

Proof of Theorem 32

Let NSRA hold at $\bar{x} \in \overline{M}$.

(1) For $\bar{y} \in M(\bar{x})$, we claim that $(\bar{y}, 0)$ is a nondegenerate minimizer for $Q(\bar{x})$ if Case I occurs and a strongly stable minimizer if Case II occurs (see Lemma 19). In both Cases, $(\bar{y}, 0)$ is an isolated minimizer for $Q(\bar{x})$. Moreover, from Lemma 7, the set $M(\bar{x})$ is compact. Hence, $M(\bar{x})$ is finite. Without loss of generality,

$M(\bar{x}) = \{\bar{y}_1, \ldots, \bar{y}_p, \bar{y}_{p+1}, \ldots, \bar{y}_l\}$, where for $\bar{y}_1, \ldots, \bar{y}_p$ Case I and for $\bar{y}_{p+1}, \ldots, \bar{y}_l$ Case II from the NSRA occurs.

(2) The existence of locally defined implicit functions

$$(y_i(\cdot), z_i(\cdot)), i = 1, \ldots, p \text{ and } (y_j^{\gamma}(\cdot), z_j^{\gamma}(\cdot)), \gamma \in \mathscr{E}(\bar{x}, \bar{y}_j), j = p+1, \ldots, l$$

with (a), (b), and (c) is from the standard results on parametric nonlinear optimization problems (recall also Lemma 18).

From Lemma 19, we also obtain the locally defined implicit functions

$$(y_j(\cdot), z_j(\cdot)), j = p+1, \ldots, l \text{ such that}$$

$(y_j, z_j)(\bar{x}) = (\bar{y}_j, 0)$,
$(y_j(x), z_j(x))$ is the locally unique local minimizer of $Q(x)$,

and

the mapping $x \mapsto (y_j(x), z_j(x))$ is continuous.

Thus, we obtain that, locally around $\bar{x}$, the set $\overline{M}$ can be described as

$$\left\{ x \middle| \begin{array}{l} z_i(x) \geq 0, \, i = 1, \ldots, p \\ z_j(x) \geq 0, \, j = p+1, \ldots, l \end{array} \right\}. \tag{3.60}$$

For the desired description (3.51), it is sufficient to prove that locally around $\bar{x}$ the following nontrivial representation is valid:

$$z_j(x) = \max_{\gamma \in \mathscr{E}(\bar{x}, \bar{y}_j)} z_j^{\gamma}(x), \, j = p+1, \ldots, l. \tag{3.61}$$

Let $j \in \{j = p+1, \ldots, l\}$ be arbitrary but fixed.

(i) $z_j(x) \geq \max\limits_{\gamma \in \mathscr{E}(\bar{x}, \bar{y}_j)} z_j^{\gamma}(x)$ since $K_+(\gamma) \subset K_0(\bar{x}, \bar{y})$ for all $\gamma \in \mathscr{E}(\bar{x}, \bar{y}_j)$.

(ii) For $z_j(x) \leq \max\limits_{\gamma \in \mathscr{E}(\bar{x}, \bar{y}_j)} z_j^{\gamma}(x)$, we find a vertex $\bar{\gamma} \in \mathscr{E}(\bar{x}, \bar{y}_j)$ (in general depending on x) with $z_j(x) = z_j^{\bar{\gamma}}(x)$.

From Lemma 20, there exists $\delta \in \Delta(\bar{x}, \bar{y})$ with $K_+(\delta) \subset K_0(x)$. We obtain the existence of a vertex $\bar{\gamma} \in \mathscr{E}(\bar{x}, \bar{y})$ such that $K_+(\bar{\gamma}) \subset K_+(\delta)$ and hence $K_+(\bar{\gamma}) \subset K_0(x)$.

Furthermore, the vectors $\begin{pmatrix} -D_y^T g_k(\bar{x}, \bar{y}) \\ 1 \end{pmatrix}$, $k \in K_+(\bar{\gamma})$, are linearly independent (see Lemma 18 (a)). Since $y_j(\cdot)$ depends continuously on x, the vectors

$$\begin{pmatrix} -D_y^T g_k(x, y_j(x)) \\ 1 \end{pmatrix}, k \in K_+(\bar{\gamma}),$$

are also linearly independent for x sufficiently close to $\bar{x}$. Moreover (and this is crucial here), they form a basis for $\mathbb{R}^m \times \mathbb{R}$ due to the fact that $|K_+(\bar{\gamma})| = m+1$ in Case II. Hence, we write

$$\begin{pmatrix} 0 \\ 1 \end{pmatrix} = \sum_{k \in K_+(\bar{\gamma})} \gamma_k(x) \begin{pmatrix} -D_y^T g_k(x, y_j(x)) \\ 1 \end{pmatrix} \text{ with } \gamma_k(x) \in \mathbb{R}, k \in K_+(\bar{\gamma}). \quad (3.62)$$

From (3.62), we see that $(y_j(x), z_j(x))$ is a critical point of the following optimization problem with equality constraints only:

$$Q^{\bar{y}}_=(x): \quad \min_{(y,z) \in \mathbb{R}^m \times \mathbb{R}} z \quad \text{s.t.} \quad z - g_k(x,y) = 0, k \in K_+(\bar{\gamma}).$$

The feasibility of $(y_j(x), z_j(x))$ for $Q^{\bar{y}}_=(x)$ is provided by $K_+(\bar{\gamma}) \subset K_0(x)$. The vector of Lagrange multipliers for $(y_j(x), z_j(x))$ is given by $\gamma(x)$.

As in Lemma 18, it can be seen that $(\bar{y}, 0)$ is a nondegenerate critical point of $Q^{\bar{y}}_=(\bar{x})$ with unique positive Lagrange multipliers $\bar{\gamma}_k, k \in K_+(\bar{\gamma})$. Thus, $\gamma_k(\bar{x}) = \bar{\gamma}_k > 0, k \in K_+(\bar{\gamma})$, and $\gamma_k(\cdot)$ depends at least continuously on x. We obtain that $\gamma_k(x), k \in K_+(\bar{\gamma})$ in (3.62) are positive for x sufficiently close to $\bar{x}$. This means that $(y_j(x), z_j(x))$ is a KKT point for $Q^{\bar{y}}(x)$. However, $(y_j^{\bar{y}}(x), z_j^{\bar{y}}(x))$ is the locally unique KKT point for $Q^{\bar{y}}(x)$ from (ii)(b). Hence, $z_j(x) = z_j^{\bar{y}}(x)$.$\square$

For the proof of Theorem 33, in particular inequality (3.65), we need some results on the geometry of the polytope $\Delta(\bar{x}, \bar{y})$ given in Lemmas 21 and 22.

Lemma 21 (Dimension at a vertex of $\Delta(\bar{x}, \bar{y})$). *Let $\bar{x} \in \overline{M}$, $\bar{y} \in M(\bar{x})$, and $\bar{\gamma}$ be a vertex of $\Delta(\bar{x}, \bar{y})$. Then,*

$$dim\left\{ span\left\{ (D_y^T g_k(\bar{x}, \bar{y}), k \in K_+(\bar{\gamma})\right\}\right\} = |K_+(\bar{\gamma})| - 1.$$

Proof. As a direct consequence of part (a) in the proof of Lemma 18 we first obtain that the vectors $(D_y^T g_k(\bar{x}, \bar{y})), k \in K_+(\bar{\gamma})$ are affine independent:

$$\text{if } \sum_{k \in K_+(\bar{\gamma})} \beta_k D_y g_k(\bar{x}, \bar{y}) = 0 \text{ and } \sum_{k \in K_+(\bar{\gamma})} \beta_k = 0, \text{ then } \beta_k = 0, k \in K_+(\bar{\gamma}). \quad (3.63)$$

Furthermore, assume that for all $K \subset K_+(\bar{\gamma})$ with $|K| = |K_+(\bar{\gamma})| - 1$ the vectors $(D_y^T g_k(\bar{x}, \bar{y})), k \in K$ are linearly dependent. There is $j \in K_+(\bar{\gamma})$ such that $\bar{\gamma}_j \neq 0$. Setting $K_j := K_+(\bar{\gamma}) \setminus \{j\}$, we get

$$\sum_{k \in K_j} \alpha_k D_y g_k(\bar{x}, \bar{y}) = 0 \text{ with } \alpha_k \in \mathbb{R} \text{ (not all vanishing)}.$$

We set $a := \sum_{k \in K_j} \alpha_k \neq 0$ due to the affine independence in (3.63). Thus,

$$\bar{\gamma}_j D_y g_j(\bar{x}, \bar{y}) + \sum_{k \in K_j} \left(\bar{\gamma}_k - \frac{\alpha_k}{a}\right) D_y g_k(\bar{x}, \bar{y}) = 0 \text{ and } \bar{\gamma}_j + \sum_{k \in K_j} \left(\bar{\gamma}_k - \frac{\alpha_k}{a}\right) = 0.$$

Expression (3.63) provides, in particular, $\bar{\gamma}_j = 0$ and hence a contradiction. $\square$

Lemma 22 (Number of vertices of the simplex $\Delta(\bar{x},\bar{y})$). *Let $\bar{x} \in \overline{M}$ and $\bar{y} \in M(\bar{x})$. Then, $\Delta(\bar{x},\bar{y})$ is a simplex. Moreover, let $\bar{\gamma}$ be a vertex of $\Delta(\bar{x},\bar{y})$. Then,*

$$|\mathscr{E}(\bar{x},\bar{y})| \leq |K_0| - |K_+(\bar{\gamma})| + 1.$$

Proof. Note that the vectors $\begin{pmatrix} -D_y^T g_k(\bar{x},\bar{y}) \\ 1 \end{pmatrix}$, $k \in K_+(\gamma)$, are linearly independent (see part (a) in the proof of Lemma 18). Hence, $\Delta(\bar{x},\bar{y})$ lies in the affine subspace

$$\left\{ \gamma \in \mathbb{R}^{|K_0(\bar{x},\bar{y})|} \,\middle|\, \sum_{k \in K_0(\bar{x},\bar{y})} \gamma_k \begin{pmatrix} -D_y^T g_k(\bar{x},\bar{y}) \\ 1 \end{pmatrix} = \begin{pmatrix} 0 \\ 1 \end{pmatrix} \right\}$$

of dimension at most $|K_0| - |K_+(\bar{\gamma})|$. In fact, $\Delta(\bar{x},\bar{y})$ is the intersection of this affine subspace and the simplex

$$\left\{ \gamma \in \mathbb{R}^{|K_0(\bar{x},\bar{y})|} \,\middle|\, \sum_{k \in K_0(\bar{x},\bar{y})} \gamma_k = 1, \ \gamma_k \geq 0, \ k \in K_0(\bar{x},\bar{y}) \right\}.$$

Consequently, $\Delta(\bar{x},\bar{y})$ is a simplex itself, and the inequality on the number of vertices $\mathscr{E}(\bar{x},\bar{y})$ of $\Delta(\bar{x},\bar{y})$ follows. $\square$

Proof of Theorem 33: (a) We prove that $\mathscr{F}$ is C_s^2-dense in $\mathscr{A}$.
We use the ideas from [38] to prove that $\mathscr{F}$ is C_s^2-generic. The proof is based on an application of the structured jet transversality theorem; for details, see, for example, [33, 63] and Section B.2.

We consider a KKT point $x \in \overline{M}$ and write $M(x) = \{y_1,\ldots,y_l\}$. There exist $v_i \in V(x,y_i)$ and $\mu_i \geq 0, i = 1,\ldots,l$ such that $Df(x) = \sum_{i=1}^{l} \mu_i v_i$. We represent any of $v_i \in V(x,y_i)$ by means of a linear combination with strictly positive multipliers of a minimal number of vectors $v_{i,j}$, $j = 1,\ldots,q_i$ forming vertices of the polytope $V(x,y_i)$. Note that

$$v_{i,j} = \sum_{k \in K_0(x,y_i)} \gamma_k^j D_x g_k(x,y_i) \text{ with a vertex } \gamma^j \in \mathscr{E}(x,y_i). \tag{3.64}$$

For $q_i, r_i \in \mathbb{N}$ and index sets $K_i, Q_{i,j} \subset K_i, i = 1,\ldots,l, j = 1,\ldots,q_i$, we consider the set Γ of $(x,y_1,\ldots,y_l,v_{1,1},\ldots,v_{l,q_l})$ such that the following conditions are satisfied:

(i) $x \in \mathbb{R}^n$, $y_i \in \mathbb{R}^m$ (pairwise different),
      $v_{i,j} \in \mathbb{R}^n$ (uniquely determined as vertices of $V(x,y_i)$ with (3.64)),
(ii) $K_i = K_0(x,y_i)$,
(iii) span$\{D_y g_k(x,y_i), k \in K_i\}$ has dimension r_i,
(iv) $Q_{i,j} = K_+(\gamma^j)$ and span$\{D_y g_k(x,y_i), k \in Q_{i,j}\}$ has dimension $|Q_{i,j}| - 1$ (see Lemma 21). Moreover, $(v_{i,j},0) \in$ span$\{Dg_k(x,y_i), k \in Q_{i,j}\}$,
(v) $Df(x) \in$ span$\{v_{i,j} \mid i = 1,\ldots,l, j = 1,\ldots,q_i\}$.

Γ constitutes a stratified manifold. Generically (from the structured jet transversality theorem), its dimension coincides with the difference between the amount of available degrees of freedom and the number of independent equations representing (i)–(v). Setting $q := \sum_{i=1}^{l} q_i$, we see that the ambient space of Γ has dimension $nq + ml + n$. Now, we count the loss of freedom $Loss_i$ caused by (ii)–(iv):

(i) nq_i, since $v_{i,j}$ are uniquely determined,
(ii) $|K_i|$,
(iii) $(m - r_i)(|K_i| - r_i)$, since $r_i \leq m$ and $r_i < |K_i|$,
(iv) $(r_i - (|Q_{i,j}| - 1))(|Q_{i,j}| - (|Q_{i,j}| - 1)) = r_i + 1 - |Q_{i,j}|$.

Hence,

$$Loss_i = nq_i + |K_i| + (m - r_i)(|K_i| - r_i) + \sum_{j=1}^{q_i} (r_i + 1 - |Q_{i,j}|).$$

Setting $M_i := \max_{j=1,\ldots,q_i} |Q_{i,j}|$, we get (from Lemma 22)

$$|K_i| \geq M_i + |\mathscr{E}(x, y_i)| - 1 \geq M_i + q_i - 1 \text{ and also } |Q_{i,j}| \leq M_i. \tag{3.65}$$

For estimating $Loss_i$ by means of (3.65), we distinguish two cases.

Case (a) $r_i = m$:

$$\begin{aligned}
Loss_i &= nq_i + |K_i| + \sum_{j=1}^{q_i} (m + 1 - |Q_{i,j}|) \\
&\geq nq_i + (M_i + q_i - 1) + mq_i + q_i - M_i q_i \\
&= nq_i + (q_i - 1)(m + 1 - M_i) + m + q_i \geq nq_i + m + q_i.
\end{aligned} \tag{3.66}$$

In the last inequality, we have $q_i - 1 \geq 0$ due to $\Delta(x, y_i) \neq \emptyset$ and also $m + 1 - M_i \geq 0$ due to Lemma 21.

Case (b) $r_i < m$: Setting for (convenience) $\beta_i := |K_i| - r_i - 1 \geq 0$,

$$\begin{aligned}
Loss_i &= nq_i + |K_i| + (m - r_i)(|K_i| - r_i) + \sum_{j=1}^{q_i} (r_i + 1 - |Q_{i,j}|) \\
&\geq nq_i + (M_i + q_i - 1) + (m - r_i)(1 + \beta_i) + r_i q_i + q_i - M_i q_i \\
&= nq_i + m + q_i + (m - r_i)\beta_i + (q_i - 1)(r_i + 1 - M_i) \geq nq_i + m + q_i.
\end{aligned} \tag{3.67}$$

In the last inequality, we have $m - r_i \geq 0$, $q_i - 1 \geq 0$, and also $r_i + 1 - M_i \geq 0$ from Lemma 21.

Finally, (v) reduces the freedom by $n - d$ degrees, where

$$d := \dim \{ \operatorname{span} \{ v_{i,j} \mid i = 1, \ldots, l, \ j = 1, \ldots, q_i \} \} \leq q.$$

Summing up over $i = 1, \ldots, l$, (i)–(v) cause a loss of at least

$$\sum_{i=1}^{l} (nq_i + m + q_i) + n - d = nq + ms + q + n - d \geq nq + mq + n \qquad (3.68)$$

degrees of freedom. This means that, for C_s^2-generic defining functions, the set Γ has dimension at most 0. Moreover, another loss of freedom would cause Γ to be empty. Thus, all inequalities in (3.65), (3.66), (3.67), and (3.68) turn to equalities to avoid the emptiness of Γ.

The equalities in (3.65) read

$$|K_i| = M_i + q_i - 1, \ |\mathscr{E}(x, y_i)| = q_i, \ \text{and} \ |Q_{i,j}| = M_i. \qquad (3.69)$$

We consider Cases (a) and (b) again, letting (3.66) and (3.67) turn to equalities.

Case (a) $r_i = m$ with equalities: Here, we get in addition to (3.69) that $q_i = 1$ or $M_i = m + 1$. If $q_i = 1$, then $\Delta(x, y_i)$ is a singleton and $|K_i| = M_i = |Q_{i,1}|$. Hence, $K_i = Q_{i,1} = K_+(\gamma_i)$, and we see that the LICQ and SC in the Case I from the NSRA hold. Moreover, the violation of the SOSC reduces the freedom and can be generically avoided here. Case I from the NSRA occurs.

If $q_i \neq 1$, then $\Delta(x, y_i)$ is not a singleton. From $|\mathscr{E}(x, y_i)| = q_i$, we see that for every vertex $\gamma \in \mathscr{E}(x, y_i)$ there exists $j \in \{1, \ldots, q_i\}$ with $Q_{i,j} = K_+(\gamma)$. Moreover, $|K_+(\gamma)| = |Q_{i,j}| = M_i = m + 1$. Case II from the NSRA occurs. We point out that in this case $\Delta(\bar{x}, \bar{y})$ is a $(|K_i| - m - 1)$-dimensional simplex in $\mathbb{R}^{|K_i| - m}$ and hence with $|K_i| - m$ vertices.

Case (b) $r_i < m$ with equalities: Here, we get in addition to (3.69) that $|K_i| = r_i + 1$ and $[q_i = 1$ or $M_i = r_i + 1]$. If $q_i = 1$, then Case I from the NSRA occurs (as above in Case (a)). If $q_i \neq 1$, then $M_i = r_i + 1$. Together with $|K_i| = r_i + 1$, we obtain $|K_i| = M_i$, and hence from (3.69) that $q_i = |K_i| - M_i + 1 = 1$, a contradiction.

From this, we claim that either Case I or Case II from the NSRA occurs; meaning ND1 in Definition 27 is generically fulfilled.

We examine the generic validity of ND2, ND3, and ND4 in Definition 27. The equality in (3.68) means that $d = q$ and hence the vectors

$$v_{i,j}, \ i = 1, \ldots, l, \ j = 1, \ldots, q_i \qquad (3.70)$$

are linearly independent. We see that ND2 is generically fulfilled. Furthermore, if $Df(x)$ does not belong to the relative interior of the cone generated by the vectors from (3.70), this causes additional loss of freedom in (v). Thus, ND3 is generically fulfilled. Moreover, the possible singularity of the matrix from ND4 also reduces the freedom and hence can be generically avoided. Finally, every KKT point x is shown to be generically nondegenerate. $\square$

Remark 26. From the proof above we see that in Case II from NSRA the Lagrange polytope $\Delta(\bar{x}, \bar{y})$ is generically a full-dimensional simplex. Here, the full-dimensionality refers to the fact that $\Delta(\bar{x}, \bar{y})$ is a $(|K_0(\bar{x}, \bar{y})| - m - 1)$-dimensional simplex, and hence with exactly $|K_0(\bar{x}, \bar{y})| - m$ vertices $\mathscr{E}(\bar{x}, \bar{y})$. Recalling Definition 28, we have under the NSRA at $\bar{x}$

$$\text{GSIP-index} = \text{QI} + \sum_{j=p+1}^{l} \left[\left| \tilde{K}_0(\tilde{x}, \tilde{y}_j) \right| - m - 1 \right].$$

Proof of Theorem 33: (b) We prove that $\mathscr{F}$ is C_s^2-open in $\mathscr{A}$.

(1) **Local argument.** First, we construct for every nondegenerate KKT point $\bar{x} \in \overline{M}$ a system of equations whose locally unique zero corresponds exactly to $\bar{x}$. We show that such a system of equations is stable in the sense that the usual implicit function theorem can be applied to follow KKT points w.r.t. local C^2-perturbations of the defining functions. Moreover, these KKT points for perturbed problems remain nondegenerate and locally unique.

To avoid unnecessary technicalities, we assume that there is only one implicit constraint in the local description (3.51) of $\overline{M}$: either $z(\cdot)$ (Case I in the NSRA) or $\max_{\gamma \in \mathscr{E}(\bar{x}, \bar{y})} z^\gamma(\cdot)$ (Case II in the NSRA). Here, we set $M(\bar{x}) = \{\bar{y}\}$.

Case I occurs for $\bar{y}$. Let $\bar{x} \in \overline{M}$ be a nondegenerate KKT point with Lagrange multiplier $\bar{\lambda} > 0$ as in Definition (27). Without loss of generality, we assume that $\bar{\lambda} = 1$. Since Case I occurs in $\bar{y}$, let $\Delta(\bar{x}, \bar{y}) = \{\bar{\gamma}\}$. We set $K_0 := K_0(\bar{x}, \bar{y})$.

We consider the following mapping $\mathscr{T} : \mathbb{R}^{n+m+1+|K_0|+1} \longrightarrow \mathbb{R}^{n+m+1+|K_0|+1}$ locally at its zero $(\bar{x}, \bar{y}, 0, \bar{\gamma}, \bar{\lambda})$:

$$\mathscr{T}(x, y, z, \gamma, \lambda) := \begin{pmatrix} D_x f(x) - \lambda \sum_{k \in K_0} \gamma_k D_x g_k(x, y) \\ \sum_{k \in K_0} \gamma_k D_y g_k(x, y) \\ \sum_{k \in K_0} \gamma_k - 1 \\ z - g_k(x, y), \, k \in K_0 \\ \sum_{k \in K_0} \gamma_k g_k(x, y) \end{pmatrix}.$$

Note that $\sum_{k \in K_0} \gamma_k g_k(x, y) = 0$ ensures feasibility for the reduced problem (3.51). In fact, let $(y(x), z(x))$ be the local minimizer of $Q(x)$. We obtain

$$\sum_{k \in K_0} \gamma_k g_k(x, y(x)) = \sum_{k \in K_0} \gamma_k z(x) = z(x) = 0.$$

Furthermore, we prove that $D\mathscr{T}(\bar{x}, \bar{y}, 0, \bar{\gamma}, \bar{\lambda})$ is nonsingular.

$$D\mathscr{T}(\bar{x}, \bar{y}, 0, \bar{\gamma}, \bar{\lambda}) = \begin{pmatrix} A & B & D \\ B^T & C & 0 \\ D^T & 0 & 0 \end{pmatrix},$$

where

$$A = D_x^2 f(\bar{x}) - \bar{\lambda} \sum_{k \in K_0} \bar{\gamma}_k D_{xx}^2 g_k(\bar{x}, \bar{y}),$$

$$B = \left(-\bar{\lambda} \sum_{k \in K_0} \bar{\gamma}_k D_{xy} g_k(\bar{x}, \bar{y}) \middle| 0 \middle| -\bar{\lambda} D_x^T g_k, k \in K_0 \right),$$

$$C = \begin{pmatrix} \sum_{k \in K_0} \bar{\gamma}_k D_{yy}^2 g_k(\bar{x}, \bar{y}) & 0 & -D_y^T g_k(\bar{x}, \bar{y}), k \in K_0 \\ 0 & 0 & 1 \\ -D_y g_k(\bar{x}, \bar{y}), k \in K_0 & 1 & 0 \end{pmatrix},$$

$$D = - \sum_{k \in K_0} \bar{\gamma}_k D_x^T g_k.$$

Note that C is nonsingular since $(\bar{y}, 0)$ is a nondegenerate minimizer of $Q(\bar{x})$ (see Theorem 2.3.2, [65]). The Schur complement of the submatrix C in $\begin{pmatrix} A & B \\ B^T & C \end{pmatrix}$ is the Hessian $D^2 L(\bar{x})$ of the Lagrange function (3.54) for the reduced problem (see Definition 27 and Remark 25). From ND4, $D^2 L(\bar{x})$ is nonsingular on the tangent space (3.55). However, the columns of D span the orthogonal complement of (3.55) due to ND1. It provides, again by means of Theorem 2.3.2 of [65], that the matrix $\begin{pmatrix} D^2 L(\bar{x}) & D \\ D^T & 0 \end{pmatrix}$ is nonsingular and hence also $D\mathscr{T}(\bar{x}, \bar{y}, 0, \bar{\gamma}, \bar{\lambda})$.

Case II occurs for $\bar{y}$. Let $\bar{x} \in \overline{M}$ be a nondegenerate KKT point with the vector of Lagrange multipliers $(\bar{\lambda}^\gamma, \gamma \in \mathscr{E}(\bar{x}, \bar{y}))$ as in Definition (27). We put $K_\gamma := K_+(\gamma)$ and $\mathscr{E} := \mathscr{E}(\bar{x}, \bar{y})$. For $\Gamma = (\Gamma_k^\gamma, k \in K_\gamma, \gamma \in \mathscr{E}) \in \mathbb{R}^{|\mathscr{E}| \cdot (m+1)}$, we set $\bar{\Gamma} := (\gamma_k, k \in K_\gamma, \gamma \in \mathscr{E})$. Moreover, $Y := (y^\gamma, \gamma \in \mathscr{E}) \in \mathbb{R}^{m \cdot |\mathscr{E}|}$, $Z := (z^\gamma, \gamma \in \mathscr{E}) \in \mathbb{R}^{|\mathscr{E}|}$, and $\Lambda := (\lambda^\gamma, \gamma \in \mathscr{E}) \in \mathbb{R}^{|\mathscr{E}|}$.

We consider the mapping

$$\mathfrak{R} : \mathbb{R}^{n + m \cdot |\mathscr{E}| + |\mathscr{E}| + |\mathscr{E}| \cdot (m+1) + |\mathscr{E}|} \longrightarrow \mathbb{R}^{n + m \cdot |\mathscr{E}| + |\mathscr{E}| + |\mathscr{E}| \cdot (m+1) + |\mathscr{E}|}$$

locally at its zero $(\bar{x}, \bar{Y}, 0, \bar{\Gamma}, \bar{\Lambda})$:

$$\mathfrak{R}(x, Y, Z, \Gamma, \Lambda) := \begin{pmatrix} D_x f(x) - \sum_{\gamma \in \mathscr{E}} \lambda^\gamma \sum_{k \in K_\gamma} \Gamma_k^\gamma D_x g_k(x, y^\gamma) \\ \sum_{k \in K_\gamma} \Gamma_k^\gamma D_y g_k(x, y^\gamma), \gamma \in \mathscr{E} \\ \sum_{k \in K_\gamma} \Gamma_k^\gamma - 1, \gamma \in \mathscr{E} \\ z^\gamma - g_k(x, y^\gamma), k \in K_\gamma, \gamma \in \mathscr{E} \\ \sum_{k \in K_\gamma} \Gamma_k^\gamma g_k(x, y^\gamma) \end{pmatrix}.$$

Note that $\sum_{k\in K_\gamma} \Gamma_k^\gamma g_k(x,y^\gamma) = 0$ ensures feasibility for the reduced problem (3.51). In fact, let $(y^\gamma(x),z^\gamma(x))$ be the local minimizers of $Q^\gamma(x)$. We obtain

$$\sum_{k\in K_\gamma} \Gamma_k^\gamma g_k(x,y^\gamma(x)) = \sum_{k\in K_\gamma} \Gamma_k^\gamma z^\gamma(x) = z^\gamma(x) = 0.$$

The nonsingularity of $D\mathfrak{R}(\bar{x},\bar{Y},0,\bar{\Gamma},\bar{\Lambda})$ can be proved analogously to Case I. To obtain the corresponding Schur complement, one uses the fact that $(\bar{y},0)$ is a non-degenerate minimizer of $Q^\gamma(\bar{x})$ (see Lemma 18). This Schur complement is exactly the Hessian $D^2L(\bar{x})$ of the Lagrange function (3.54) for the reduced problem (see Definition 27). Here, the formula for the second derivative of the implicit constraint $z^\gamma(\cdot)$ is used (see Remark 25). Finally, ND1 and ND4 imply the nonsingularity of $D\mathfrak{R}(\bar{x},\bar{Y},0,\bar{\Gamma},\bar{\Lambda})$.

In both Cases I and II, the implicit function theorem can be applied to follow KKT points w.r.t. local C^2-perturbations of the defining functions. By means of continuity arguments in ND1–ND4, these KKT points for perturbed problems remain nondegenerate and locally unique.

(2) **Global argument.** The global argument is standard. We only stress that under Assumption B the set-valued mapping $(x,g) \rightrightarrows M_g(x)$ is upper semicontinuous w.r.t. the topology in $\mathbb{R}^n \times \left[C^2(\mathbb{R}^n \times \mathbb{R}^m)\right]^{k+1}$. Hence, $M_g(x)$ is locally bounded w.r.t. C^2-perturbations of the defining functions (see also Lemma 7). The global issue is due to the strong C_s^2-topology. $\square$

Remark 27. The proof above provides a description of nondegenerate KKT points as solutions of certain stable equations involving first- and second-order information on the defining functions. (We refer to mappings $\mathscr{T}$ and $\mathfrak{R}$ from the proof of Theorem 33). This fact might be used to establish some (nonsmooth) versions of the Newton method for GSIPs (see [12, 81, 112]). This issue is a topic of current research.

Application to SIP

We consider the special case of the standard SIP characterized by a constant set $Y := Y(x)$,

$$\text{SIP:} \quad \text{minimize } f(x) \text{ s.t. } x \in M, \tag{3.71}$$

with

$$M := \{x \in \mathbb{R}^n \,|\, g_0(x,y) \geq 0 \text{ for all } y \in Y\}$$

and a compact set

$$Y := \{y \in \mathbb{R}^m \,|\, g_k(y) \leq 0, k = 1,\dots,s\}.$$

For $\bar{x} \in M$, we denote by $E_{g_0}(\bar{x})$ the active index set $\{y \in Y \,|\, g_0(\bar{x},y) = 0\}$. It consists of all global minimizers for $g_0(\bar{x},\cdot)|_Y$.

Let the Sym-MFCQ be fulfilled for the SIP to provide the description

$$M = \{x \in \mathbb{R}^n \mid \sigma(x,y) \geq 0 \text{ for all } y \in \mathbb{R}^m\}.$$

The Sym-MFCQ is equivalent here to the well-known extended Mangasarian-Fromovitz constraint qualification (EMFCQ) for the SIP and the standard MFCQ for Y at all $y \in E_{g_0}(\bar{x})$ (see Lemma 13).

The well-known reduction ansatz for SIP (cf. [37]) states that all $y \in E_{g_0}(\bar{x})$ are nondegenerate minimizers for $g_0(\bar{x}, \cdot)_{|Y}$. It turns out that the reduction ansatz for SIP corresponds exactly to Case I in NRSA for GSIPs. The proof of Theorem 34 is straightforward.

Theorem 34 (Reduction ansatz for SIP vs. Case I in NSRA). *Let M be given as in (3.71) and $\bar{x} \in M$. Then, $\bar{y} \in E_{g_0}(\bar{x})$ is a nondegenerate minimizer for $g_0(\bar{x}, \cdot)_{|Y}$ if and only if $(\bar{y}, 0)$ with $\bar{y} \in M(\bar{x})$ is a nondegenerate minimizer for $Q(\bar{x})$.*

From Theorem 34, we see that, compared with the SIP, a main new issue for the GSIP is Case II in NRSA.

Remark 28. It is worth mentioning that the well-known corresponding genericity results for KKT points in SIPthe for the reduced feasible set (i.e., ND2–ND4) can be achieved by means of the perturbations of $g_0(\cdot, \cdot)$ only. The set Y remains unchanged.

3.4 Critical point theory

Under the Sym-MFCQ, two basic theorems from Morse theory are proved (see Section 1.2 and Section A.1). Outside the set of Karush-Kuhn-Tucker (KKT) points, continuous deformation of lower-level sets can be performed. As a consequence, the topological data (such as the number of connected components) then remain invariant. However, when passing a KKT level, the topology of the lower-level set changes via the attachment of a q-dimensional cell. The dimension q equals the so-called GSIP-index of the (nondegenerate) KKT point. Here, the nonsmooth symmetric reduction ansatz (NSRA) allows us to perform a local reduction of the GSIP to a disjunctive optimization problem. The GSIP-index then coincides with the stationary index from the corresponding disjunctive optimization problem. We refer the reader to [74] for details; see also [116].

Deformation and cell attachment

For $a, b \in \mathbb{R}$, $a < b$, define the sets

$$M^a := \{x \in M \mid f(x) \leq a\}, \overline{M}^a := \{x \in \overline{M} \mid f(x) \leq a\},$$

$$\overline{M}_a^b := \{x \in \overline{M} \mid a < f(x) < b\},$$
$$\mathrm{lev}(\overline{M}, a) := \{x \in \overline{M} \mid f(x) = a\}.$$

Our aim is to prove the following result.

Theorem 35 (Critical point theory on M). *Let condition CC be fulfilled, the Sym-MFCQ hold at all points $x \in M^{max}$ and M be bounded. Then, the following results are valid:*

(a)   **(Deformation theorem on M)** *If $\overline{M}_a^b$ does not contain any KKT point, then M^a is homotopy-equivalent to M^b.*

(b)   **(Cell-attachment theorem on M)** *If $\overline{M}_a^b$ contains exactly one nondegenerate KKT point, say $\bar{x}$, and if $a < f(\bar{x}) < b$ and the GSIP-index of $\bar{x}$ is equal to q, then M^b is homotopy-equivalent to M^a with a q-cell attached.*

We describe the idea behind the proof of the main deformation and cell-attachment theorem. For that, we consider an explicit description of $\overline{M}$. In Theorem 30, it is shown that under the compactness condition (CC) and the symmetric Mangasarian-Fromovitz constraint qualification (Sym-MFCQ) (see Definitions 18 and 23) the closure of the feasible set is given by

$$\overline{M} = M^{max}, \tag{3.72}$$

where

$$M^{max} = \left\{ x \in \mathbb{R}^n \mid \max_{0 \le k \le s} g_k(x, y) \ge 0 \text{ for all } y \in \mathbb{R}^m \right\}.$$

Having description (3.72) in mind, we consider the relaxed problem

$$\overline{GSIP}: \quad \text{minimize } f(x) \text{ s.t. } x \in M^{max}. \tag{3.73}$$

We prove the corresponding deformation and cell-attachment results for $\overline{GSIP}$.

Theorem 36 (Critical point theory on $\overline{M}$). *Let condition CC be fulfilled, the Sym-MFCQ hold at all points $x \in M^{max}$ and M be bounded. Then, the following results are valid:*

(a)   **(Deformation theorem on $\overline{M}$)** *If $\overline{M}_a^b$ does not contain KKT points, then $\overline{M}^a$ is a strong deformation retract of $\overline{M}^b$.*

(b)   **(Cell-attachment theorem on $\overline{M}$)** *If $\overline{M}_a^b$ contains exactly one nondegenerate KKT point, say $\bar{x}$, and if $a < f(\bar{x}) < b$ and the GSIP-index of $\bar{x}$ is equal to q, then $\overline{M}^b$ is homotopy-equivalent to $\overline{M}^a$ with a q-cell attached.*

For proving Theorem 36, the Sym-MFCQ becomes crucial. Moreover, we use the fact that under condition CC and the Sym-MFCQ the set $\overline{M}$ is a Lipschitz manifold (see Theorem 28). Furthermore, we link the topology of the lower-level set for the GSIP and $\overline{GSIP}$, respectively.

Theorem 37 (Topology of M^a vs. $\overline{M}^a$). *If the set $\mathrm{lev}(\overline{M},a)$ does not contain any KKT points (i.e., the level $a \in \mathbb{R}$ is regular), M^a is homotopy-equivalent to $\overline{M}^a$.*

Finally, the main Theorem 35 follows easily from both of these results.

Proofs of main results

For the proof of Theorem 36, we need the following simple but crucial lemma.

Lemma 23 (Local descriptions of $M^{\mathbf{max}}$ and φ). *Let condition CC be fulfilled, and let $\bar{x} \in M^{max}$. Then, there exist some neighborhood $U_{\bar{x}}$ of $\bar{x}$ and a nonempty compact set $W \subset \mathbb{R}^m$ such that*

$$M^{max} \cap U_{\bar{x}} = \{x \in U_{\bar{x}} \mid \sigma(x,y) \geq 0 \text{ for all } y \in W\}$$

and $\varphi(x) = \min_{y \in W} \sigma(x,y)$, $x \in U_{\bar{x}}$.
If additionally $\varphi(x) = 0$, then $M(x) = \{y \in W \mid \sigma(x,y) = 0\}$, $x \in U_{\bar{x}}$.

Proof. From Lemma 15, condition CC implies that

(C1)   for all $x \in \mathbb{R}^n$ and sequences $(y_k)_{k \in \mathbb{N}} \subset \mathbb{R}^m$ with $\sigma(x,y_k) \to \varphi(x)$,
        $k \to \infty$, there exists a convergent subsequence of $(y_k)_{k \in \mathbb{N}}$ and
(C2)   the mapping $x \rightrightarrows M^{\varphi}(x) := \{y \in \mathbb{R}^m \mid \sigma(x,y) = \varphi(x)\}$ is locally bounded;
        that is, for all $\bar{x} \in \mathbb{R}^n$ there exists an open neighborhood $U_{\bar{x}} \subseteq \mathbb{R}^n$ of $\bar{x}$ such
        that $\bigcup_{x \in U_{\bar{x}}} M^{\varphi}(x)$ is bounded.

Let $\bar{x} \in M^{max}$. Using the neighborhood $U_{\bar{x}} \subset \mathbb{R}^n$ of $\bar{x}$ from (C2), we set

$$W := \overline{\bigcup_{x \in U_{\bar{x}}} M^{\varphi}(x)},$$

where W is a compact set to (C2).

Now, let $(y_k)_{k \in \mathbb{N}}$ be a minimizing sequence for $\sigma(x, \cdot)$ with $x \in U_{\bar{x}}$,

$$\sigma(x,y_k) \to \varphi(x), \quad k \to \infty.$$

(C1) implies the existence of a subsequence $(y_{k_l})_{l \in \mathbb{N}}$ of $(y_k)_{k \in \mathbb{N}}$ with

$$y_{k_l} \to \bar{y} \in \mathbb{R}^m, \quad l \to \infty.$$

From the continuity of $\sigma(x, \cdot)$, we have

$$\sigma(x,y_{k_l}) \to \sigma(x,\bar{y}), \quad l \to \infty.$$

Since $(y_{k_l})_{l \in \mathbb{N}}$ is a minimizing sequence, it holds that

$$\varphi(x) = \sigma(x,\bar{y}).$$

Thus, we get $\bar{y} \in M^{\varphi}(x) \subseteq W$. Note that for $x \in U_{\bar{r}}$ with $\varphi(x) = 0$ it holds that $M^{\psi}(x) = M(x)$.□

Proof of Theorem 36

(a) We show that $\overline{M}^a$ is a strong deformation retract of $\overline{M}^b$ in several steps.

Step 1: Existence of Sym-MFCQ vectors $\xi_{\bar{x}}$, $\bar{x} \in \overline{M}_a^b$.

Since every $\bar{x} \in \overline{M}_a^b$ is not a KKT point, we obtain

$$0 \notin \text{conv}\left(\{-D^T f(\bar{x})\} \cup V(\bar{x})\right). \tag{3.74}$$

Due to the compactness of $V(\bar{x})$ (which easily follows from the compactness of $M(\bar{x})$; see Lemma 23), a separation argument can be used as in [110]. From (3.74), we obtain the existence of a vector $\xi_{\bar{x}} \in \mathbb{R}^n$ such that

$$D^T f(\bar{x}) \cdot \xi_{\bar{x}} < 0$$

and

$$v \cdot \xi_{\bar{x}} > 0 \text{ for all } v \in V(\bar{x}).$$

The latter means, in particular, that $\xi_{\bar{x}}$ is a Sym-MFCQ vector at $\bar{x}$.

Step 2: Localization of Sym-MFCQ vectors $\xi_{\bar{x}}$, $\bar{x} \in \overline{M}_a^b$.

Let $\bar{x} \in \overline{M}_a^b$. We claim that there exists an open neighborhood $O_{\bar{x}}$ of $\bar{x}$ such that

$$D^T f(x) \cdot \xi_{\bar{x}} < 0$$

and

$$v \cdot \xi_{\bar{x}} > 0 \text{ for all } v \in V(x), x \in O_{\bar{x}} \cap \overline{M}. \tag{3.75}$$

We refer the reader to Theorem 22 for details on the proof of (3.75). It follows mainly from the fact that the mappings $x \rightrightarrows M(x)$ and $(x, y) \rightrightarrows V(x, y)$ are upper-semicontinuous due to the local representation of $\overline{M}$ from Lemma 23. Moreover, $M(\bar{x})$ and $V(\bar{x})$ are compact sets.

Step 3: Globalization of Sym-MFCQ vectors $\xi_{\bar{x}}$, $\bar{x} \in \overline{M}_a^b$.

Due to the boundedness of M, $\overline{M}_a^b$ is a compact set. Hence, we get from the open covering $\left\{O_{\bar{x}} \mid \bar{x} \in \overline{M}_a^b\right\}$ of $\overline{M}_a^b$ a subcovering $\{O_{\bar{x}_j} \mid \bar{x}_j \in \overline{M}_a^b, j \in J\}$ with a finite set J. Using a C^{∞}-partition of unity $\{\phi_j\}$ subordinate to $\{O_{\bar{x}_j} \mid \bar{x}_j \in \overline{M}_a^b, j \in J\}$ we define with $\xi_{\bar{x}_j}$ from Step 2 a C^{∞}-vector field

$$\xi(x) := \sum_{j \in J} \phi_j(x) \xi_{\bar{x}_j} \text{ for } x \in \overline{M}_a^b.$$

The last induces a flow $\Psi(t, \cdot)$ on $\overline{M}_a^b$ (see Theorem 3.3.14 of [63] for details). Since $\xi(x)$ is a convex combination of the vectors $\{\xi_{\bar{x}_j} \mid x \in O_{\bar{x}_j}, j \in J\}$, we obtain

$$D^T f(x) \cdot \xi(x) < 0$$

and

$$v \cdot \xi(x) > 0 \text{ for all } v \in V(x). \tag{3.76}$$

Step 4: Feasibility and descent behavior of $\Psi(t, \cdot)$ on $\overline{M}_a^b$.
Our aim is to show that there exist $\varepsilon > 0$ and $\bar{t} > 0$ such that

$$\Psi(t, x) \in \overline{M} \text{ for all } t \in [0, \bar{t}], x \in \overline{M}_a^b,$$

and

$$\Psi(\bar{t}, x) \in \overline{M}^{f(x)-\varepsilon} \text{ for all } x \in \overline{M}_a^b.$$

Step 4a: Feasibility of $\Psi(t, \cdot)$ on $\overline{M}_a^b$.
For $\bar{x} \in \overline{M}_a^b$, we consider the local description of $\overline{M}$ with $U_{\bar{x}}$ and W as given in Lemma 23. By shrinking $U_{\bar{x}}$, we may assume that it is a compact neighborhood of $\bar{x}$.

First, we show the local feasibility of $\Psi(t, \cdot)$ on $\overline{M}_a^b \cap U_{\bar{x}}$; namely, that there exists $t_{\bar{x}} > 0$ such that

$$\Psi(t, x) \in \overline{M} \text{ for all } t \in [0, t_{\bar{x}}], x \in \overline{M}_a^b \cap U_{\bar{x}}. \tag{3.77}$$

Let $x \in \overline{M}_a^b \cap U_{\bar{x}}$ and $y \in W$.
   Case 1: $y \notin M(x)$. Then, $\max\limits_{0 \le k \le s} g_k(x, y) > 0$.
   Case 2: $y \in M(x)$. We write the Taylor expansion for $g_k(\Psi(\cdot, x), y)$, $k \in K_0(\bar{x}, y)$ at 0:

$$g_k(\Psi(t, x), y) = t \left[D_x g_k(x, y) \cdot \xi(x) + \frac{o_k(t, x, y)}{t} \right]. \tag{3.78}$$

We choose a vector $v \in V(x, y) \subset V(x)$, written as

$$v = \sum_{k \in K_0(x, y)} \gamma^k(x, y) D_x g_k(x, y) \text{ with}$$

$$\sum_{k \in K_0(x, y)} \gamma^k(x, y) D_y g_k(x, y) = 0, \quad \sum_{k \in K_0(x, y)} \gamma^k(x, y) = 1, \gamma^k(x, y) \ge 0.$$

Multiplying (3.78) by $\gamma_k(x, y)$ and summing up, we obtain

$$\sum_{k \in K_0(x, y)} \gamma^k(x, y) g_k(\Psi(t, x), y) = t \left[v \cdot \xi(x) + \sum_{k \in K_0(x, y)} \gamma^k(x, y) \frac{o_k(t, x, y)}{t} \right] \ge$$

$$t \left[\min_{v \in V(x)} v \cdot \xi(x) + \sum_{k \in K_0(x, y)} \gamma^k(x, y) \frac{o_k(t, x, y)}{t} \right] \ge$$

$$t \left[\inf_{x \in U_{\bar{x}}} \min_{v \in V(x)} v \cdot \xi(x) + \sum_{k \in K_0(x, y)} \gamma^k(x, y) \frac{o_k(t, x, y)}{t} \right].$$

Hence, we get

$$\max_{k=0,\ldots,s} g_k(\Psi(t,x),y) \geq \max_{k\in K_0(x,y)} g_k(\Psi(t,x),y) >$$

$$\sum_{k\in K_0(x,y)} \gamma^k(x,y) \max_{k\in K_0(x,y)} g_k(\Psi(t,x),y) \geq \sum_{k\in K_0(x,y)} \gamma^k(x,y) g_k(\Psi(t,x),y) \geq$$

$$t\left[\inf_{x\in U_{\bar{x}}} \min_{v\in V(x)} v\cdot\xi(x) + \sum_{k\in K_0(x,y)} \gamma^k(x,y)\frac{o_k(t,x,y)}{t}\right].$$

We claim that

$$\inf_{x\in U_{\bar{x}}} \min_{v\in V(x)} v\cdot\xi(x) > 0. \tag{3.79}$$

In fact, the description of $M(x)$ in Lemma 23 easily provides that $V(x)$ is compact. Hence, the set $\bigcup_{x\in U_{\bar{x}}} V(x)$ is also compact. Finally, for the validity of (3.79), it is crucial to note that $\xi(\cdot)$ is continuous (see Steps 2 and 3).

Moreover,

$$\sum_{k\in K_0(x,y)} \gamma^k(x,y)\frac{o_k(t,x,y)}{t} \longrightarrow 0 (\text{as } t\longrightarrow 0)$$

$$\text{uniformly on } (x,y)\in U_{\bar{x}}\times \overline{\bigcup_{x\in U_{\bar{x}}} M(x)}.$$

The latter comes from the fact that

$$U_{\bar{x}}\times \overline{\bigcup_{x\in U_{\bar{x}}} M(x)} \subset U_{\bar{x}}\times W \text{ is compact,} \quad \sum_{k\in K_0(x,y)} \gamma^k(x,y) = 1, \text{ and}$$

$$\frac{o_k(t,x,y)}{t} = \int_0^1 [D_x g_k(\Psi(st,x),y) - D_x g_k(x,y)]\cdot\xi(x)\,ds$$

$$\text{is continuous w.r.t. } (t,x,y).$$

From all of this, we obtain the existence of a real number $t_{\bar{x}} > 0$ (which is independent from $x\in U_{\bar{x}}$) such that (3.77) holds.

Now, it is straightforward to see that there exists $\bar{t}_1 > 0$ such that

$$\Psi(t,x)\in \overline{M} \text{ for all } t\in[0,\bar{t}_1], x\in \overline{M}_a^b.$$

In fact, we consider the covering of a compact set $\overline{M}_a^b$ by the compact neighborhoods $\left\{U_{\bar{x}}\,|\,\bar{x}\in \overline{M}_a^b\right\}$. Then, using a subcovering $\{U_{\bar{x}_i}\,|\,\bar{x}_i\in \overline{M}_a^b, i\in I\}$ with a finite set I, we define

$$\bar{t}_1 := \min_{i\in I}\{t_{\bar{x}_i}\} > 0.$$

Step 4b: Descent behavior of $\Psi(t,\cdot)$ on $\overline{M}_a^b$.
We show that there exist $\varepsilon > 0$ and $\bar{t}_2 > 0$ such that

$$\Psi(\bar{t}_2, x) \in \overline{M}^{f(x)-\varepsilon} \text{ for all } x \in \overline{M}_a^b. \tag{3.80}$$

Having proved the feasibility of $\Psi(t, \cdot)$ on $\overline{M}_a^b$ in Step 4a, we only need to consider the function $f(\Psi(\cdot, x))$ for $x \in \overline{M}_a^b$.

We write the Taylor expansion for $f(\Psi(\cdot, x))$ at 0:

$$f(\Psi(t, x)) = f(x) + t \left[Df(x) \cdot \xi(x) + \frac{o(t, x)}{t} \right].$$

We get

$$f(\Psi(t, x)) \leq f(x) + t \left[\sup_{x \in \overline{M}_a^b} Df(x) \cdot \xi(x) + \frac{o(t, x)}{t} \right].$$

From (3.76) and the continuity of $\xi(\cdot)$, we have

$$\sup_{x \in \overline{M}_a^b} Df(x) \cdot \xi(x) = \max_{x \in \overline{M}_a^b} Df(x) \cdot \xi(x) < 0.$$

Moreover,

$$\frac{o(t, x)}{t} \longrightarrow 0 (\text{as } t \longrightarrow 0) \text{ uniformly on } \overline{M}_a^b.$$

The latter comes from the fact that $\overline{M}_a^b$ is compact and

$$\frac{o(t, x)}{t} = \int_0^1 [Df(\Psi(st, x)) - Df(x)] \cdot \xi(x) \, ds \text{ is continuous w.r.t. } (t, x).$$

Thus, we conclude that real numbers $\varepsilon > 0$ and $\bar{t}_2 > 0$ exist such that (3.80) holds.

From this, we obtain the validity of Step 4 putting $\bar{t} := \min\{\bar{t}_1, \bar{t}_2\}$.

Step 5: Deformation via Ψ.

From Step 4, we obtain for $x \in \overline{M}_a^b$ a unique $t_a(x) \geq 0$ with $\Psi(t_a(x), x) \in \overline{M}^a$. It is not hard (but technical) to realize that $t_a : x \longrightarrow t_a(x)$ is Lipschitz. This follows mainly from the application of the standard implicit function theorem and the fact that $\overline{M}_a^b$ is a Lipschitz manifold (see Theorem 28).

Finally, we define $r : [0, 1] \times \overline{M}^b \longrightarrow \overline{M}^b$ as

$$r(\tau, x) := \begin{cases} x & \text{for } x \in \overline{M}^a, \ \tau \in [0, 1] \\ \Psi(\tau t_a(x), x) & \text{for } x \in \overline{M}_a^b, \ \tau \in [0, 1]. \end{cases}$$

The mapping r provides that $\overline{M}^a$ is a strong deformation retract of $\overline{M}^b$.

(b) By virtue of the Deformation part (a) and Theorem 32, the proof of the cell-attachment theorem becomes standard. In fact, the deformation theorem allows deformations up to an arbitrarily small neighborhood U of the nondegenerate KKT point $\bar{x}$. In such a neighborhood U we can use Theorem 32 to apply the corresponding cell-attachment theorem for disjunctive optimization problems (see Theorem

3.2 in [71]). Hence, we obtain that $\overline{M}^b \cap U$ is homotopy-equivalent to $\overline{M}^a \cap U$ with a q-cell attached. This provides the validity of (b). $\square$

Proof of Theorem 37

For $\delta > 0$, we set

$$\overline{M}^a(\delta) := \left\{ x \in \overline{M} \mid f(x) \le a, \, \varphi(x) \ge \delta \right\}$$

and

$$\left(\overline{M}^a\right)_0^\delta := \left\{ x \in \overline{M} \mid f(x) \le a, \, 0 \le \varphi(x) \le \delta \right\}.$$

First, we prove that there exists a real number $\bar{\delta} > 0$ such that

$$\overline{M}^a(\bar{\delta}) \text{ is a strong deformation retract of } \overline{M}^a. \tag{3.81}$$

Since $\mathrm{lev}(\overline{M}, a)$ does not contain any KKT points, we obtain from the continuity and compactness arguments that there exists $\bar{\delta} > 0$ such that every $\bar{x} \in \mathrm{lev}(\overline{M}, a) \cap \left(\overline{M}^a\right)_0^\delta$ is not a KKT point for the optimization problem

$$\overline{\mathrm{GSIP}}(\bar{x}): \quad \text{minimize } f(x) \text{ s.t. } x \in \overline{M}(\bar{x})$$

with

$$\overline{M}(\bar{x}) := \left\{ x \in \mathbb{R}^n \mid \varphi(x) - \varphi(\bar{x}) \ge 0 \right\}.$$

Note that $\overline{M}(\bar{x})$ is the feasible set of $\overline{\mathrm{GSIP}}$ w.r.t. the perturbed data functions $g_k^{\bar{x}}(x, y) := g_k(x, y) - \varphi(\bar{x})$. We write $V^{\bar{x}}(\cdot)$ for the corresponding set $V(\cdot)$ as in (3.4) w.r.t. $g^{\bar{x}}(\cdot, \cdot)$.

Thus, for $\bar{x} \in \mathrm{lev}(\overline{M}, a) \cap \left(\overline{M}^a\right)_\delta^0$, we obtain the existence of a vector $\xi_{\bar{x}} \in \mathbb{R}^n$ such that

$$D^T f(\bar{x}) \cdot \xi_{\bar{x}} < 0$$

and

$$v \cdot \xi_{\bar{x}} > 0 \text{ for all } v \in V^{\bar{x}}(\bar{x}).$$

Moreover, for $\bar{x} \in \left(\overline{M}^a\right)_0^\delta$ with $f(x) < a$, we also obtain the existence of a vector $\xi_{\bar{x}} \in \mathbb{R}^n$ such that

$$v \cdot \xi_{\bar{x}} > 0 \text{ for all } v \in V^{\bar{x}}(\bar{x}).$$

This is due to the stability property of the Sym-MFCQ w.r.t. C_s^1-topology (see Theorem 22). Note that C_s^1-topology coincides with the usual C^1-topology on the compact set $\overline{M}$. Here, $\bar{\delta}$ can be taken smaller if needed.

Furthermore, we proceed analogously to the proof of Theorem 36(b) to construct a C^∞-flow $\Psi(t, \cdot)$ on $\left(\overline{M}^a\right)_0^\delta$ such that

$$\Psi(t, x) \in \overline{M}^a \text{ for all } t \in [0, \bar{t}], \, x \in \left(\overline{M}^a\right)_0^\delta$$

and

$$\Psi(\bar{t}, x) \in \overline{M}^a(\varphi(x) + \varepsilon) \text{ for all } x \in \left(\overline{M}^a\right)_0^\delta$$

with some $\varepsilon > 0$ and $\bar{t} > 0$.

As in the proof of Theorem 36(b), we see that the flow Ψ induces a strong retraction mapping between $\overline{M}^a(\bar{\delta})$ and $\overline{M}^a$.

We claim that the same flow Ψ induces a strong retraction mapping between $\overline{M}^a(\bar{\delta})$ and M^a itself. In fact, from the estimation of o-terms in the proof of Theorem 36(b), we see that

$$\Psi(t,x) \in \text{int}(\overline{M}) \text{ for all } t \in (0,\bar{t}], x \in \left(\overline{M}^a\right)_0^\delta.$$

Moreover, Theorem 28 implies that $\text{int}(\overline{M}) = \text{int}M$. Hence,

$$\Psi(t,x) \in \text{int}(M) \text{ for all } t \in (0,\bar{t}], x \in \left(\overline{M}^a\right)_0^\delta.$$

This shows that $\overline{M}^a(\bar{\delta})$ is a strong deformation retract of M^a.

Altogether, we have that $\overline{M}^a(\bar{\delta})$ is a strong deformation retract of both $\overline{M}^a$ and M^a. Consequently, $\overline{M}^a(\bar{\delta})$ is homotopy-equivalent to $\overline{M}^a$ and M^a. Since homotopy-equivalence is an equivalence relation, we obtain that $\overline{M}^a$ and M^a are homotopy-equivalent.$\square$

Proof of Theorem 35

The assertions follow directly from Theorems 36 and 37. We only note that the cell attachment on a homotopy-equivalent space is induced via the corresponding homotopy mapping. $\square$

Chapter 4
Mathematical Programming Problems with Vanishing Constraints

We study mathematical programming problems with vanishing constraints (MPVC) from the topological point of view. The critical point theory for MPVCs is presented. For that, we introduce the notion of a T-stationary point for the MPVC.

4.1 Applications and examples

We consider the mathematical programming problem with vanishing constraints (MPVC)

$$\text{MPVC:} \quad \min f(x) \text{ s.t. } x \in M[h,g,H,G], \tag{4.1}$$

with

$$M[h,g,H,G] := \{x \in \mathbb{R}^n \mid H_m(x) \geq 0, \, H_m(x)G_m(x) \leq 0, \, m = 1,\dots,k,$$
$$h_i(x) = 0, \, i \in I, \, g_j(x) \geq 0, \, j \in J\},$$

where $h := (h_i, i \in I)^T \in C^2(\mathbb{R}^n, \mathbb{R}^{|I|})$, $g := (g_j, j \in J)^T \in C^2(\mathbb{R}^n, \mathbb{R}^{|J|})$, $H :=$ $(H_m, m = 1,\dots,k)^T$, $G := (G_m, m = 1,\dots,k)^T \in C^2(\mathbb{R}^n, \mathbb{R}^k)$, $f \in C^2(\mathbb{R}^n, \mathbb{R})$, $|I| \leq n$, $k \geq 0$, $|J| < \infty$. For simplicity, we write M for $M[h,g,H,G]$ if no confusion is possible.

The MPVC was introduced in [1] as a model for structural and topology optimization. It is motivated by the fact that the constraint G_m does not play any role whenever H_m is active. We refer the reader to [43, 44, 45, 46, 56, 55] for more details on optimality conditions, constraint qualifications, sensitivity, and numerical methods for the MPVC. Note that additional constraints $G_m(x) \geq 0$, $m = 1,\dots,k$ would restrict the MPVC to a so-called mathematical program with complementarity constraints (MPCC). In addition to an MPCC feasible set, M is glued together from manifold pieces of **different dimensions** along their strata. Indeed, a typical MPVC feasible set

$$\mathbb{V} := \{(x,y) \mid x \geq 0, \, xy \geq 0\}$$

is depicted in Figure 21.

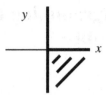

Figure 21 $\mathbb{V}$ **solution set of the basic vanishing constraint relation**

It represents the solution set of the basic vanishing constraint relations and exhibits one- and two-dimensional parts glued together at $(0,0)$.

Truss topology optimization

The following application of truss topology optimization is from [1]. The problem is to construct the optimal design of a truss structure. Let us consider a set of potential bars that are defined by the coordinates of their end nodes. For each potential bar, material parameters are given (Young's modulus E_i, relative moment of inertia s_i, and stress bounds $\sigma_i^t > 0$ and $\sigma_i^c < 0$ for tension and compression, respectively). These parameters are needed for the formulation of constraints to prevent structural failure when the potential bar is realized as a real bar. This the case if the calculated cross-sectional area a_i is positive. Finally, boundary conditions and external loads at some of the nodes are given. The problem now is to find cross-sectional areas a_i for each potential bar such that failure of the whole structure is prevented, the external load is carried by the structure, and a suitable objective function is minimal. The latter is usually the total weight of the structure or its deformation energy. In view of a practical realization of the calculated structure after optimization, one hopes that the optimal design will make use of only a few of the potential bars. Such behavior is typical in applied truss topology optimization problems. The main difficulty in formulating (and solving) the problem lies in the fact that constraints on structural failure can be formulated in a well-defined way only if there is some material giving mechanical response. However, most potential bars will possess a zero cross section at the optimizer. Hence, the truss topology optimization problem might be formulated as an MPVC:

$$\text{Truss-Top:} \quad \underset{(a,u)\in\mathbb{R}^M\times\mathbb{R}^d}{\text{minimize}} \ f(a,u) \ \text{s.t.}$$

$$g(a,u) \le 0, \ K(a)u = f^{\text{ext}},$$

$$a_i \ge 0, \ i = 1,\dots,M,$$

$$\sigma_i^c \le \sigma(a,u) \le \sigma_i^t \ \text{if} \ a_i > 0, \ i = 1,\dots,M,$$

$$f_i^{\text{int}}(a,u) \ge f_i^{\text{buck}}(a,u) \ \text{if} \ a_i > 0, \ i = 1,\dots,M.$$

Here, the vector $a \in \mathbb{R}^M$ contains the vector of cross-sectional areas of the potential bars and $u \in \mathbb{R}^d$ denotes the vector of nodal displacements of the structure under load, where d is the so-called degree of freedom of the structure, the number of free nodal displacement coordinates. The state variable u serves as an auxiliary variable. The objective function f expresses structural weight. The nonlinear system of equations $K(a)u = f^{\text{ext}}$ symbolizes force equilibrium of (given) external loads $f^{\text{ext}} \in \mathbb{R}^d$ and internal forces expressed via Hooke's law in terms of displacements and cross sections. The matrix $K(a) \in \mathbb{R}^{d \times d}$ is the global stiffness matrix corresponding to the structure a. The constraint $g(a, u) \leq 0$ is a resource constraint. If $a_i > 0$, then $\sigma_i(a, u) \in \mathbb{R}$ is the stress along the i-th bar. Similarly, if $a_i > 0$, $f_i^{\text{int}}(a, u) \in \mathbb{R}$ denotes the internal force along the i-th bar, and $f_i^{\text{buck}}(a)$ corresponds to the permitted Euler buckling force. Then the constraints on stresses and on local buckling make sense only if $a_i > 0$. Therefore, they must vanish from the problem if $a_i = 0$.

4.2 Critical point theory

Our goal is the investigation of the MPVC from a topological point of view. To this end, we introduce the new notion of a T-stationary point for the MPVC (see Definition 29). It turns out that the concept of T-stationarity is an adequate stationarity concept for topological considerations. In fact, we introduce the letter "T" for a stationarity concept that is **topologically** relevant rather than giving a tight first-order condition for local minimizers (see also the discussion below).

Furthermore, we study the behavior of the topological properties of lower-level sets

$$M^a := \{x \in M \,|\, f(x) \leq a\}$$

for the MPVC as the level $a \in \mathbb{R}$ varies. In particular, within this context, we present two basic theorems from Morse theory (see [63, 93] and Section A.1). First, we show that, for $a < b$, the set M^a is a strong deformation retract of M^b if the (compact) set

$$M_a^b := \{x \in M \,|\, a \leq f(x) \leq b\}$$

does not contain T-stationary points (see Theorem 40(a)). Second, if M_a^b contains exactly one (nondegenerate) T-stationary point, then M^b is shown to be homotopy-equivalent to M^a with a q-cell attached (see Theorem 40(b)). Here, the dimension q is the T-index (see Definitions 29 and 31). We refer the reader to [20] for details.

T-stationarity

Given $\bar{x} \in M$, we define the (active) index sets

$$J_0 = J_0(\bar{x}) := \{j \in J \,|\, g_j(\bar{x}) = 0\},$$

$$I_{0+} = I_{0+}(\bar{x}) := \{m \in \{1, \ldots k\} \mid H_m(\bar{x}) = 0, G_m(\bar{x}) > 0\},$$

$$I_{0-} = I_{0-}(\bar{x}) := \{m \in \{1, \ldots k\} \mid H_m(\bar{x}) = 0, G_m(\bar{x}) < 0\},$$

$$I_{+0} = I_{+0}(\bar{x}) := \{m \in \{1, \ldots k\} \mid H_m(\bar{x}) > 0, G_m(\bar{x}) = 0\},$$

$$I_{00} = I_{00}(\bar{x}) := \{m \in \{1, \ldots k\} \mid H_m(\bar{x}) = 0, G_m(\bar{x}) = 0\}.$$

We call $J_0(\bar{x})$ the active inequality index set and $I_{00}(\bar{x})$ the biactive index set at $\bar{x}$. Note that, locally around $\bar{x}$, for $m \in I_{0+}$, the function H_m behaves like an ordinary equality constraint ($H_m(x) = 0$). For $m \in I_{0-}$ or $m \in I_{+0}$, the functions H_m and G_m behave locally like inequality constraints ($H_m(x) \geq 0$ or $G_m(x) \leq 0$, respectively).

Furthermore, we recall the well-known linear independence constraint qualification (LICQ) for the MPVC (e.g., [1]), which is said to hold at $\bar{x} \in M$ if the vectors

$$D^T h_i(\bar{x}), i \in I, D^T H_m(\bar{x}), m \in I_{0+},$$
$$D^T g_j(\bar{x}), j \in J_0, D^T H_m(\bar{x}), m \in I_{0-}, D^T G_m(\bar{x}), m \in I_{+0},$$
$$D^T H_m(\bar{x}), D^T G_m(\bar{x}), m \in I_{00}$$

are linearly independent.

We introduce the notion of a T-stationary point, which is crucial for the following.

Definition 29 (T-stationary point). A point $\bar{x} \in M$ is called T-stationary for the MPVC if there exist real numbers $\bar{\lambda}_i, i \in I, \bar{\alpha}_m, m \in I_{0+}, \bar{\mu}_j, j \in J_0, \bar{\beta}_m, m \in I_{0-},$ $\bar{\gamma}_m, m \in I_{+0}, \bar{\delta}_m^H, \bar{\delta}_m^G, m \in I_{00}$ (Lagrange multipliers) such that

$$Df(\bar{x}) = \sum_{i \in I} \bar{\lambda}_i Dh_i(\bar{x}) + \sum_{m \in I_{0+}} \bar{\alpha}_m DH_m(\bar{x})$$

$$+ \sum_{j \in J_0} \bar{\mu}_j Dg_j(\bar{x}) + \sum_{m \in I_{0-}} \bar{\beta}_m DH_m(\bar{x}) + \sum_{m \in I_{+0}} \bar{\gamma}_m DG_m(\bar{x})$$

$$+ \sum_{m \in I_{00}} \left(\bar{\delta}_m^H DH_m(\bar{x}) + \bar{\delta}_m^G DG_m(\bar{x}) \right), \qquad (4.2)$$

$$\bar{\mu}_j \geq 0 \text{ for all } j \in J_0, \qquad (4.3)$$

$$\bar{\beta}_m \geq 0 \text{ for all } m \in I_{0-}, \qquad (4.4)$$

$$\bar{\gamma}_m \leq 0 \text{ for all } m \in I_{+0}, \qquad (4.5)$$

$$\bar{\delta}_m^G \leq 0 \text{ and } \bar{\delta}_m^H \cdot \bar{\delta}_m^G \geq 0 \text{ for all } m \in I_{00}. \qquad (4.6)$$

In the case where the LICQ holds at $\bar{x} \in M$, the Lagrange multipliers in (4.2) are uniquely determined.

Given a T-stationary point $\bar{x} \in M$ for the MPVC, we set

$$M(\bar{x}) := \{x \in \mathbb{R}^n \mid h_i(x) = 0, i \in I, H_m(x) = 0, m \in I_{0+}, g_j(x) = 0, j \in J_0,$$
$$H_m(x) = 0, m \in I_{0-}, G_m(x) = 0, m \in I_{+0},$$
$$H_m(x) = 0, G_m(x) = 0, m \in I_{00}\}.$$

Obviously, $M(\bar{x}) \subset M$ and, in the case where the LICQ holds at $\bar{x}$, $M(\bar{x})$ is locally at $\bar{x}$ a C^2-manifold.

Definition 30 (Nondegenerate T-stationary point). A T-stationary point $\bar{x} \in M$ with Lagrange multipliers as in Definition 29 is called nondegenerate if the following conditions are satisfied:

ND1: LICQ holds at $\bar{x}$.
ND2: $\bar{\mu}_j > 0$ for all $j \in J_0$, $\bar{\beta}_m > 0$ for all $m \in I_{0-}$, $\bar{\gamma}_m < 0$ for all $m \in I_{+0}$.
ND3: $D^2 L(\bar{x}) \mid_{T_{\bar{x}}M(\bar{x})}$ is nonsingular.
ND4: $\bar{\delta}_m^H < 0$ and $\bar{\delta}_m^G < 0$ for all $m \in I_{00}$.

Here, the matrix $D^2 L$ stands for the Hessian of the Lagrange function L,

$$L(x) := f(x) - \sum_{i \in I} \bar{\lambda}_i h_i(x) - \sum_{m \in I_{0+}} \bar{\alpha}_m H_m(x)$$

$$- \sum_{j \in J_0} \bar{\mu}_j g_j(x) - \sum_{m \in I_{0-}} \bar{\beta}_m H_m(x) - \sum_{m \in I_{+0}} \bar{\gamma}_m G_m(x)$$

$$- \sum_{m \in I_{00}} \left(\bar{\delta}_m^H H_m(x) - \bar{\delta}_m^G G_m(x) \right), \tag{4.7}$$

and $T_{\bar{x}}M(\bar{x})$ denotes the tangent space of $M(\bar{x})$ at $\bar{x}$,

$$\begin{aligned} T_{\bar{x}}M(\bar{x}) := \{\xi \in \mathbb{R}^n \mid &Dh_i(\bar{x})\xi = 0, i \in I, \\ &DH_m(\bar{x})\xi = 0, m \in I_{0+}, \\ &Dg_j(\bar{x})\xi = 0, j \in J_0, \\ &DH_m(\bar{x})\xi = 0, m \in I_{0-}, \\ &DG_m(\bar{x})\xi = 0, m \in I_{+0}, \\ &DH_m(\bar{x})\xi = 0, DG_m(\bar{x})\xi = 0, m \in I_{00}\}. \end{aligned}$$

Condition ND3 means that the matrix $V^T D^2 L(\bar{x}) V$ is nonsingular, where V is some matrix whose columns form a basis for the tangent space $T_{\bar{x}}M(\bar{x})$.

Definition 31 (T-index). Let $\bar{x} \in M$ be a nondegenerate T-stationary point with Lagrange multipliers as in Definition 30. The number of negative eigenvalues of $D^2 L(\bar{x}) \mid_{T_{\bar{x}}M(\bar{x})}$ in ND3 is called the quadratic index (QI) of $\bar{x}$. The number of negative pairs $(\bar{\delta}_m^H, \bar{\delta}_m^G)$, $m \in I_{00}$ in ND4 equals $|I_{00}|$ and is called the biactive index (BI) of $\bar{x}$. The number $(QI + BI)$ is called the T-index of $\bar{x}$.

Note that in the absence of biactive vanishing constraints, the T-index has only the QI part and coincides with the well-known quadratic index of a nondegenerate Karush-Kuhn-Tucker point in nonlinear programming or, equivalently, with the Morse index (see [63, 83, 93] and Section 1.4). Also note that the biactive index BI is completely determined by the cardinality of I_{00}, in contrast to, for example, the biactive index for MPCCs as defined in Section 2.3 (see also [69]).

The following proposition uses the T-index for the characterization of a local minimizer.

Proposition 8. *(i)   Assume that $\bar{x}$ is a local minimizer for the MPVC and that the LICQ holds at $\bar{x}$. Then, $\bar{x}$ is a T-stationary point for the MPVC.*

(ii)   Let $\bar{x}$ be a nondegenerate T-stationary point for the MPVC. Then, $\bar{x}$ is a local minimizer for the MPVC if and only if its T-index is equal to zero.

Proof. (i) From [1] it is known that under the LICQ a local minimizer $\bar{x}$ for the MPVC is a strongly stationary point, meaning (4.2)–(4.5) hold and

$$\bar{\delta}_m^G = 0 \text{ and } \bar{\delta}_m^H \geq 0 \text{ for all } m \in I_{00}. \tag{4.8}$$

Clearly, a strongly stationary point is T-stationary as well.

(ii) Let $\bar{x}$ be a nondegenerate T-stationary local minimizer for the MPVC. As in (i), we claim that $\bar{x}$ is also strongly stationary. Comparing ND4 and (4.8), we see that $BI = |I_{00}| = 0$. Then, locally around $\bar{x}$, the MPVC behave like an ordinary nonlinear program, and using standard results on the quadratic index, we obtain that $QI = 0$. The other direction is trivial.   □

The next genericity and stability results justify the LICQ assumption as well as the introduction of nondegeneracy for T-stationary points in the MPVC.

Theorem 38 (Genericity and Stability).

(i)   Let $\mathscr{F}$ denote the subset of

$$C^2(\mathbb{R}^n, \mathbb{R}^{|I|}) \times C^2(\mathbb{R}^n, \mathbb{R}^{|J|}) \times C^2(\mathbb{R}^n, \mathbb{R}^k) \times C^2(\mathbb{R}^n, \mathbb{R}^k)$$

consisting of those (h, g, H, G) for which the LICQ holds at all points $x \in M[h, g, H, G]$. Then, $\mathscr{F}$ is C_s^2-open and -dense.

(ii)   Let $\mathscr{D}$ denote the subset of

$$C^2(\mathbb{R}^n, \mathbb{R}) \times C^2(\mathbb{R}^n, \mathbb{R}^{|I|}) \times C^2(\mathbb{R}^n, \mathbb{R}^{|J|}) \times C^2(\mathbb{R}^n, \mathbb{R}^k) \times C^2(\mathbb{R}^n, \mathbb{R}^k)$$

consisting of those problem data (f, h, g, H, G) for which each T-stationary point is nondegenerate. Then, $\mathscr{D}$ is C_s^2-open and -dense.

Proof. (i) We define the set

$$M_{\text{DISJ}} := \{x \in \mathbb{R}^n \mid \max\{H_m(x), G_m(x)\} \geq 0, m = 1, \ldots, k,$$
$$h_j(x) = 0, i \in I, g_j(x) \geq 0, j \in J\}.$$

M_{DISJ} is the feasible set of a disjunctive optimization problem (see [71]). We obtain from the corresponding results on disjunctive optimization that the subset of problem data for which the LICQ holds for all $x \in M_{\text{DISJ}}$ is C_s^2-dense and C_s^2-open (see [71], Lemmas 2.4 and 2.5). Recalling that the notions of the LICQ for disjunctive optimization problems and MPVCs are the same, and that M is a subset of M_{DISJ}, the desired result follows immediately.

(ii) The proof is based on the application of the jet transversality theorem, for details, see, for example, [63] and Section B.2. For subsets $\tilde{J} \subseteq J$ and $\tilde{H}, \tilde{G} \subseteq \{1, \ldots, k\}$, and sets $D_{\tilde{J}} \subseteq \tilde{J}$, $D_{\tilde{H}} \subseteq \tilde{H}$, and $D_{\tilde{G}} \subseteq \tilde{G}$ and $r \in \{0, \ldots, \dim(T_{\bar{x}}M(\bar{x}))\}$, we consider the set Γ of x such that the following conditions are satisfied:

(m1) $g_j(x) = H_i(x) = G_l(x) = 0$ for all $j \in \tilde{J}, i \in \tilde{H}, l \in \tilde{G}$.

(m2) $Df(x) \in \text{span} \left\{ \begin{array}{l} Dg_j(x), j \in \tilde{J} \setminus D_{\tilde{J}}, \\ DH_i(x), i \in \tilde{H} \setminus D_{\tilde{H}}, \\ DG_l(x), l \in \tilde{G} \setminus D_{\tilde{G}} \end{array} \right\}.$

(m3)   The matrix $D^2L(x)|_{T_{\bar{x}}M(\bar{x})}$ has rank r.

Now it suffices to show that Γ is generically empty whenever one of the sets $D_{\tilde{J}}$, $D_{\tilde{H}}$, or $D_{\tilde{G}}$ is nonempty or the rank r of the matrix in (m3) is not full. This would mean, respectively, that a Lagrange multiplier in the equality (4.2) vanishes (see ND2, ND4) or the rank condition ND3 fails to hold.

In fact, the available degrees of freedom of the variables involved in Γ are n. The loss of freedom caused by (m1) is at least $d := |\tilde{J}| + |\tilde{H}| + |\tilde{G}|$, and the loss of freedom caused by (m2) is at least (supposing that the gradients on the right-hand side are linearly independent (ND1) and the sets $D_{\tilde{J}}, D_{\tilde{H}}, D_{\tilde{G}}$ are empty) $n - d$. Hence, the total loss of freedom is n. We conclude that a further nondegeneracy would exceed the total available degrees of freedom n. By virtue of the jet transversality theorem, generically the set Γ must be empty.

For the openness result, we can argue in a standard way (see, for example, [63]). Locally, T-stationarity can be rewritten via stable equations. Then, the implicit function theorem for Banach spaces can be applied to follow nondegenerate T-stationary points w.r.t. (local) C^2-perturbations of defining functions. Then a standard globalization procedure exploiting the specific properties of the strong C^2-topology can be used to construct a (global) C_s^2-neighborhood of problem data for which the nondegeneracy property is stable.$\square$

Morse lemma for the MPVC

For the proof of the results mentioned above we locally describe the MPVC feasible set under the LICQ (see Lemma 24). Moreover, an equivariant Morse lemma for the MPVC is derived in order to obtain suitable normal forms for the objective function at nondegenerate T-stationary points (see Theorem 39).

Without loss of generality, we assume that at the particular point of interest $\bar{x} \in M$ it holds that

$$J_0 = \{1, \ldots, |J_0|\},$$

$$I_{0+} = \{1, \ldots, |I_{0+}|\},$$

$$I_{0-} = \{|I_{0+}| + 1, \ldots, |I_{0+}| + |I_{0-}|\},$$

$$I_{+0} = \{|I_{0+}| + |I_{0-}| + 1, \ldots, |I_{0+}| + |I_{0-}| + |I_{+0}|\},$$

$$I_{00} = \{|I_{0+}| + |I_{0-}| + |I_{+0}| + 1, \ldots, |I_{0+}| + |I_{0-}| + |I_{+0}| + |I_{00}|\}.$$

We put $s := |I| + |I_{0+}|$, $r := s + |J_0| + |I_{0-}|$, $q := r + |I_{+0}|$, $p := n - q - 2|I_{00}|$.

For the proof of Theorem 40, we need a local description of the MPVC feasible set under the LICQ.

Definition 32. The feasible set M admits a local C^r-coordinate system of $\mathbb{R}^n$ ($r \geq 1$) at $\bar{x}$ by means of a C^r-diffeomorphism $\Phi : U \longrightarrow V$ with open $\mathbb{R}^n$-neighborhoods U and V of $\bar{x}$ and 0, respectively, if it holds that

(i) $\Phi(\bar{x}) = 0$,

(ii) $\Phi(M \cap U) = \left(\{0_s\} \times \mathbb{H}^{|J_0| + |I_{0-}|} \times (-\mathbb{H})^{|I_{+0}|} \times \mathbb{V}^{|I_{00}|} \times \mathbb{R}^p \right) \cap V.$

Lemma 24. *Suppose that the LICQ holds at $\bar{x} \in M$. Then M admits a local C^2-coordinate system of $\mathbb{R}^n$ at $\bar{x}$.*

Proof. Choose vectors $\xi_l \in \mathbb{R}^n$, $l = 1, \dots, p$, which form, together with the vectors

$$D^T h_i(\bar{x}), i \in I, D^T H_m(\bar{x}), m \in I_{0+},$$
$$D^T g_j(\bar{x}), j \in J_0, D^T H_m(\bar{x}), m \in I_{0-}, D^T G_m(\bar{x}), m \in I_{+0},$$
$$D^T H_m(\bar{x}), D^T G_m(\bar{x}), m \in I_{00},$$

a basis for $\mathbb{R}^n$. Next we put

$$
\left.
\begin{aligned}
y_i &:= h_i(x), i \in I, \\
y_{|I|+m} &:= H_m(x), m \in I_{0+}, \\
y_{|I|+|I_{0+}|+j} &:= g_j(x), j \in J_0, \\
y_{|I|+|J_0|+m} &:= H_m(x), m \in I_{0-}, \\
y_{|I|+|J_0|+m} &:= G_m(x), m \in I_{+0}, \\
y_{|I|+|J_0|+2m-1} &:= H_m(x), m \in I_{00}, \\
y_{|I|+|J_0|+2m} &:= G_m(x), m \in I_{00}, \\
y_{n-p+l} &:= \xi_l^T(x - \bar{x}), l = 1, \dots, p
\end{aligned}
\right\}
\tag{4.9}
$$

or, for short,

$$y = \Phi(x). \tag{4.10}$$

Note that $\Phi \in C^2(\mathbb{R}^n, \mathbb{R}^n)$, $\Phi(\bar{x}) = 0$, and the Jacobian matrix $D\Phi(\bar{x})$ is nonsingular (by virtue of the LICQ and the choice of ξ_l, $l = 1, \dots, p$). By means of the implicit function theorem, there exist open neighborhoods U of $\bar{x}$ and V of 0 such that $\Phi : U \longrightarrow V$ is a C^2-diffeomorphism. By shrinking U if necessary, we can guarantee that $J_0(x) \subset J_0$, $I_{0-}(x) \subset I_{0-}$, $I_{+0}(x) \subset I_{+0}$ and $I_{00}(x) \subset I_{00}$ for all $x \in M \cap U$. Thus, property (ii) in Definition 32 follows directly from the definition of Φ. $\square$

Definition 33. We will refer to the C^2-diffeomorphism Φ defined by (4.9) and (4.10) as the standard diffeomorphism.

Remark 29. It follows from the proof of Lemma 24 that the Lagrange multipliers at a nondegenerate T-stationary point are the corresponding partial derivatives of the objective function in new coordinates given by the standard diffeomorphism (see [65], Lemma 2.2.1). Moreover, the Hessian with respect to the last p coordinates corresponds to the restriction of the Lagrange function's Hessian on the respective tangent space (cf. [65], Lemma 2.2.10).

We derive an equivariant Morse lemma for the MPVC in order to obtain suitable normal forms for the objective function at nondegenerate T-stationary points.

Theorem 39 (Morse lemma for MPVC). *Suppose that $\bar{x}$ is a nondegenerate T-stationary point for the MPVC with quadratic index QI, biactive index BI, and T-index = QI + BI. Then, there exists a local C^1-coordinate system $\Psi : U \longrightarrow V$ of $\mathbb{R}^n$ around $\bar{x}$ (according to Definition 32) such that*

$$f \circ \Psi^{-1}(0_s, y_{s+1}, \ldots, y_n) =$$

$$f(\bar{x}) + \sum_{i=1}^{|J_0|+|I_{0-}|} y_{i+s} - \sum_{j=1}^{|I_{+0}|} y_{j+r} - \sum_{m=1}^{|I_{00}|} \left(y_{2j-1+q} + y_{2j+q} \right) + \sum_{k=1}^{p} \pm y_{k+n-p}^2, \quad (4.11)$$

where $y \in \{0_s\} \times \mathbb{H}^{|J_0|+|I_{0-}|} \times (-\mathbb{H})^{|I_{+0}|} \times \mathbb{V}^{|I_{00}|} \times \mathbb{R}^p$. Moreover, in (4.11) there are exactly $BI = |I_{00}|$ negative linear pairs and QI negative squares.

Proof. Without loss of generality, we may assume $f(\bar{x}) = 0$. Let $\Phi : U \longrightarrow V$ be a standard diffeomorphism according to Definition 33. We put $\bar{f} := f \circ \Phi^{-1}$ on the set $\left(\{0_s\} \times \mathbb{H}^{|J_0|+|I_{0-}|} \times (-\mathbb{H})^{|I_{+0}|} \times \mathbb{V}^{|I_{00}|} \times \mathbb{R}^p \right) \cap V$. We may assume $s = 0$ from now on. In view of Remark 29, we have at the origin

(i) $\quad \dfrac{\partial \bar{f}}{\partial y_i} > 0, \, i = 1, \ldots, |J_0| + |I_{0-}|,$

(ii) $\quad \dfrac{\partial \bar{f}}{\partial y_{j+r}} < 0, \, j = 1, \ldots, |I_{+0}|,$

(iii) $\quad \dfrac{\partial \bar{f}}{\partial y_{2m-1+q}} < 0$ and $\dfrac{\partial \bar{f}}{\partial y_{2m+q}} < 0$ for exactly BI indices $m = 1, \ldots, |I_{00}|,$

(iv) $\quad \dfrac{\partial \bar{f}}{\partial y_{k+n-p}} = 0, \, k = 1, \ldots, p$ and $\left(\dfrac{\partial^2 \bar{f}}{\partial y_{k_1+n-p} \partial y_{k_2+n-p}} \right)_{1 \leq k_1, k_2 \leq p}$ is a nonsingular matrix with QI negative eigenvalues.

We denote $\bar{f}$ by f. Under the following coordinate transformations the set $\mathbb{H}^{|J_0|+|I_{0-}|} \times (-\mathbb{H})^{|I_{+0}|} \times \mathbb{V}^{|I_{00}|} \times \mathbb{R}^p$ will be transformed in itself (equivariance). As an abbreviation, we put $y = (Y_{n-p}, Y^p)$, where $Y_{n-p} = (y_1, \ldots, y_{n-p})$ and $Y^p = (y_{n-p+1}, \ldots, y_n)$. We write

$$f(Y_{n-p}, Y^p) = f(0, Y^p) + \int_0^1 \frac{d}{dt} f(t Y_{n-p}, Y^p) dt = f(0, Y^p) + \sum_{i=1}^{n-p} y_i d_i(y),$$

where $d_i \in C^1, \, i = 1, \ldots, n-p$.

In view of (iv), we may apply the Morse lemma on the C^2-function $f(0, Y^p)$ (see Theorem 2.8.2 of [63]) without affecting the coordinates Y_{n-p}. The corresponding coordinate transformation is of class C^1. Denoting the transformed functions f, d_j again by f, d_j, we obtain

$$f(y) = \sum_{i=1}^{n-p} y_i d_i(y) + \sum_{k=1}^{p} \pm y_{k+n-p}^2.$$

Note that $d_i(0) = \dfrac{\partial f}{\partial y_i}(0)$, $i = 1, \ldots, n - p$. Recalling (i)–(iii), we have

$$y_i |d_i(y)|, \; i = 1, \ldots, n - p, \quad y_j, \; j = n - p + 1, \ldots, n, \tag{4.12}$$

as new local C^1-coordinates. Denoting the transformed function f again by f and recalling the signs in (i)–(iii), we obtain (4.11). Here, the coordinate transformation Ψ is understood as the composite of all previous ones. $\square$

Deformation and Cell-Attachment

We state and prove the main deformation and cell-attachment theorems for the MPVC. Recall that for $a, b \in \mathbb{R}$, $a < b$, the sets M^a and M_a^b are defined as

$$M^a := \{ x \in M \mid f(x) \leq a \}$$

and

$$M_a^b := \{ x \in M \mid a \leq f(x) \leq b \}.$$

Theorem 40. *Let M_a^b be compact, and suppose that the LICQ is satisfied at all points $x \in M_a^b$.*

(a)   **(Deformation theorem)** *If M_a^b does not contain any T-stationary point for the MPVC, then M^a is a strong deformation retract of M^b.*

(b)   **(Cell-attachment theorem)** *If M_a^b contains exactly one (nondegenerate) T-stationary point for the MPVC, say $\bar{x}$, and if $a < f(\bar{x}) < b$ and the T-index of $\bar{x}$ is equal to q, then M^b is homotopy-equivalent to M^a with a q-cell attached.*

Proof. (a) Let $\bar{x} \in M_a^b$. After a coordinate transformation with the standard diffeomorphism from Definition 32 and Remark 29, we may assume that $\bar{x} = 0$ and locally $M = \{0_s\} \times \mathbb{H}^{|J_0| + |I_{0-}|} \times (-\mathbb{H})^{|I_{+0}|} \times \mathbb{V}^{|I_{00}|} \times \mathbb{R}^p$. From Remark 29 and the fact that $\bar{x}$ is not a T-stationary point (see Definition 29), one of the following cases holds:

(a)   There exists $j \in \{1, \ldots, p\}$ with $\dfrac{\partial f}{\partial y_{n-p+j}}(0) \neq 0$.

(b)   There exists $j \in \{1, \ldots, |J_0| + |I_{0-}|\}$ with $\dfrac{\partial f}{\partial y_{s+j}}(0) < 0$.

(c)   There exists $j \in \{1, \ldots, |I_{+0}|\}$ with $\dfrac{\partial f}{\partial y_{r+j}}(0) > 0$.

(d)   There exists $m \in I_{00}$ with $\dfrac{\partial f}{\partial y_{q+2m}}(0) > 0$.

(e)   There exists $m \in I_{00}$ with $\dfrac{\partial f}{\partial y_{q+2m-1}}(0) > 0$ and $\dfrac{\partial f}{\partial y_{q+2m}}(0) < 0$.

We set

$$D := \{ x \in M_a^b \mid \text{one of cases a)–d) holds} \} \quad \text{and} \quad L := M_a^b \setminus D.$$

The proof consists of the local argument and its globalization.

Local argument. We prove that for each $\bar{x} \in M_a^b$ there exists an $\mathbb{R}^n$-neighborhood $\mathscr{U}_{\bar{x}}$ of $\bar{x}$, a $t_{\bar{x}} > 0$, and a flow

$$\Psi^{\bar{x}} : [0, t_{\bar{x}}) \times M^b \cap \mathscr{U}_{\bar{x}} \to M, \ (t, x) \mapsto \Psi^{\bar{x}}(t, x),$$

with:

1. $\Psi^{\bar{x}}(0, x) = x$ for all $x \in M^b \cap \mathscr{U}_{\bar{x}}$.
2. $\Psi^{\bar{x}}(t_2, \Psi^{\bar{x}}(t_1, x)) = \Psi^{\bar{x}}(t_1 + t_2, x)$ for all $x \in M^b \cap \mathscr{U}_{\bar{x}}$ and $t_1, t_2 \geq 0$ with $t_1 + t_2 \in [0, t_{\bar{x}})$.
3. $f(\Psi^{\bar{x}}(t, x)) \leq f(x) - t$ for all $x \in M^b \cap \mathscr{U}_{\bar{x}}$ and $t \in [0, t_{\bar{x}})$.
4. If $\bar{x} \in D$, then $\Psi^{\bar{x}}$ is a C^2-flow corresponding to a C^1-vector field. If $\bar{x} \in L$, then $\Psi^{\bar{x}}$ is a Lipschitz flow.

We consider the constructions of the local flows in Cases a)–e).

Cases (a)–(c). We can use standard methods to construct a local flow induced by a C^1-vector field. To see this, note that the behavior of partial derivatives in Cases (a)–(c) give us a descent direction that—due to the structure of M in local coordinates—is feasible for $t_{\bar{x}} > 0$. (This is a standard construction for generalized manifolds with boundary; see Theorems 2.7.6 and 3.2.26 of [63] for details and also the proof of Theorem 20).

If the violation of T-stationarity is exclusively due to the coordinates belonging to the set $\mathbb{V}^{|I_{00}|}$ (i.e. one of the cases (d) and (e) holds), we have to construct a new flow.

Case (d). Using an (additional) local coordinate transformation leaving M invariant, analogous to the proof of Theorem 39, we obtain

$$f(y) = y_{q+2m} + f(y_1, \ldots, y_{q+2m-1}, 0, y_{q+2m+1}, \ldots, y_n).$$

We define a local vector field as $\tilde{F}^{\bar{x}}(y) := (0, \ldots, 0, -1, 0, \ldots, 0)^T$. After the inverse change of local coordinates, $\tilde{F}^{\bar{x}}$ induces the flow, which fits the local argument.

Case e). Again, as in the proof of Theorem 39, we may assume that

$$f(y) = y_{q+2m-1} - y_{q+2m} + f(y_1, \ldots, y_{q+2m-2}, 0, 0, y_{q+2m+1}, \ldots, y_n).$$

We define a two-dimensional flow $\Phi(t, z)$ for $z = (z_1, z_2) \in \mathbb{V}$ as

$$\Phi(t, z_1, z_2) := \begin{cases} \left(\begin{array}{c} \max\left\{0, \left(1 - \frac{t}{z_1 - z_2}\right) \cdot z_1\right\} \\ \left[\left(1 - \frac{t}{z_1 - z_2}\right) \cdot z_2\right]^- + [t - (z_1 - z_2)]^+ \end{array} \right) & \text{for } z_2 < 0, \\[20pt] \left(\begin{array}{c} 0 \\ t - (z_1 - z_2) \end{array} \right) & \text{for } z_2 \geq 0. \end{cases}$$

Here, $[\cdot]^-$ is the negative and $[\cdot]^+$ the positive part of a real number.

Note that the flow Φ is Lipschitz on $\mathbb{R} \times \mathbb{V}$. Moreover, due to the definition of Φ, we get that the flow $\Psi^{\bar{x}}$ defined (again in new coordinates) by

$$\Psi_i(y) := \begin{cases} y_i & \text{for } i \in \{1,\ldots,n\} \setminus \{q+2m-1, q+2m\}, \\ \Phi_1(y_{q+2m-1}) & \text{for } i = q+2m-1, \\ \Phi_2(y_{q+2m}) & \text{for } i = q+2m. \end{cases}$$

fits the local argument. Here, Ψ_i and Φ_i stands for the i-th components of Ψ and Φ, respectively.

Globalization. Now we construct a global flow Ψ on M_a^b. Suppose for a moment that there exists a flow Ψ_L on a neighborhood $\mathscr{U}_L$ of L with the properties (i) to (iv). We choose a smaller neighborhood $\mathscr{W}_L$ of L such that the closure $\overline{\mathscr{W}_L}$ of $\mathscr{W}_L$ is contained in $\mathscr{U}_L$. Furthermore, we choose an arbitrary open covering $\{\mathscr{U}_x \mid x \in M_a^b \setminus \mathscr{U}_L\}$ of $M_a^b \setminus \mathscr{U}_L$ induced by the domains of the C^2-flows corresponding to cases (a)–(d). Since $M_a^b \setminus \mathscr{U}_L$ is compact we find a finite subcovering $\{\mathscr{U}_x \mid x \in \bar{D}\}$. Here $\bar{D}$ is a finite subset of D. Without loss of generality, we may assume that for all $x \in \bar{D}$ the closure $\overline{\mathscr{U}_x}$ of $\mathscr{U}_x$ is disjoint with $\overline{\mathscr{W}_L}$. By construction, it holds that $\{\mathscr{U}_x \mid x \in \bar{D}\} \cup (\mathscr{U}_L \setminus \overline{\mathscr{W}_L})$ is a finite open covering of $M_a^b \setminus \mathscr{W}_L$. The crucial argument is now that outside the set L the flow Ψ_L is induced by a C^1-vector field. (Note that Φ only has a singularity for $t = z_1 - z_2$.) Therefore, we can construct a flow on $M_a^b \setminus \overline{\mathscr{W}_L}$ by using a C^∞-partition of unity subordinate to the open covering $\{\mathscr{U}_x \mid x \in \bar{D}\} \cup (\mathscr{U}_L \setminus \overline{\mathscr{W}_L})$. This enables us to construct a global C^1-vector field. The flow Ψ_D obtained by integration fulfills the desired properties. (See Theorem 3.3.14 of [63] for details on this procedure.) By construction, Ψ_L and Ψ_D can be glued together into one flow Ψ on M_a^b.

We obtain for $x \in M_a^b$ a unique $t_a(x) > 0$ with $\Psi(t_a(x), x) \in M^a$ from the properties of Ψ (which are induced by local properties of the flows Ψ^x). It is not hard (but technical) to realize that $t_a : x \mapsto t_a(x)$ is Lipschitz. Finally, we define $r : [0,1] \times M^b \to M^b$ as

$$r(\tau, x) := \begin{cases} x & \text{for } x \in M^a, \ \tau \in [0,1], \\ \Psi(\tau \cdot t_a(x), x) & \text{for } x \in M_a^b, \ \tau \in [0,1]. \end{cases}$$

The mapping r provides that M^a is a strong deformation retract of M^b.

It remains to construct the flow Ψ_L. Since this construction is highly technical, we only present a short outline. The main idea is to construct the flow along strata inside L; here the strata are induced by all possible subsets of active constraints $H_1, G_1, \ldots, H_m, G_m$. Along a given stratum, we find a differentiable family of standard coordinate systems (see Lemma 24). This enables us to define a flow along this stratum just by applying flows like Φ on fixed components that depend on the coordinate system. By introducing notions of a distance from a point in the embedding space to the strata, we can construct homotopies (via Lipschitz continuous time scaling) between the different branches of the stratification and the corresponding flows. (For details on such constructions with the aid of tube systems, we refer to [27].)

(b) From the deformation theorem (Theorem 40(a)), we may assume that, w.l.o.g., a and b are small enough that we can work in local coordinates. Therefore, we consider the normal form (2.19) from Theorem 39,

$$f(y) = \sum_{i=1}^{|J_0|+|I_{0-}|} y_{s+i} - \sum_{j=1}^{|I_{+0}|} y_{r+j} - \sum_{m=1}^{|I_{00}|} (y_{q+2m-1} + y_{q+2m}) + \sum_{l=1}^{p} \pm y_{n-p+l}^2,$$

with $y \in M := \{0_s\} \times \mathbb{H}^{|J_0|+|I_{0-}|} \times (-\mathbb{H})^{|I_{+0}|} \times \mathbb{V}^{|I_{00}|} \times \mathbb{R}^p$.

We set

$$M_{\text{MPCC}} := \{0_s\} \times \mathbb{H}^{|J_0|+|I_{0-}|} \times (-\mathbb{H})^{|I_{+0}|} \times (\partial \mathbb{H}^2)^{|I_{00}|} \times \mathbb{R}^p.$$

Note that M_{MPCC} differs from M by the appearance of $(\partial \mathbb{H}^2)^{|I_{00}|}$ instead of $\mathbb{V}^{|I_{00}|}$.

For $c \in \mathbb{R}$, it holds that $M_{\text{MPCC}}^c := \{y \in M_{\text{MPCC}} \mid f(y) \leq c\}$ is a strong deformation retract of $M^c := \{y \in M \mid f(y) \leq c\}$. In fact, we define a mapping $g : M^c \to M_{\text{MPCC}}^c$ with

$$y_i \mapsto \begin{cases} 0 & i \in \{q+2m \mid m = 1, \ldots, |I_{00}|\} \text{ and } y_i < 0, \\ y_i & \text{else.} \end{cases}$$

We see that there is a (convex combination) homotopy between g and the identity on M^c. If $(y_{q+2m-1}, y_{q+2m}) \in \mathbb{V}$, then $(y_{q+2m-1}, 0) \in \partial \mathbb{H}^2$ and, moreover, $f(g(y)) \leq f(y)$ for all $y \in M^c$ (i.e., g in fact maps to M_{MPCC}^c). Hence, M_{MPCC}^c is a strong deformation retract of M^c.

According to Definition 11, it holds that $\bar{y} = 0$ is a nondegenerate C-stationary point of the MPCC defined by f and the set M_{MPCC}. Since $\bar{y} = 0$ is the only C-stationary point, Theorem 20(b) implies that M_{MPCC}^b is homotopy-equivalent to M_{MPCC}^a with a $\tilde{q}$-cell attached. Note that $\tilde{q}$ is the so-called C-index for the corresponding MPCC. Here, we have that the C-index $\tilde{q}$ w.r.t. the MPCC coincides with the T-index q w.r.t. the MPVC. Hence

$$M_{\text{MPCC}}^b \simeq (M_{\text{MPCC}}^a \text{ with a } q\text{-cell attached}).$$

We know from the considerations above that M^c is homotopy-equivalent to M_{MPCC}^c for $c = a, b$. Furthermore, we note that the cell attachment on a homotopy-equivalent space is induced via the corresponding homotopy mapping. Finally, using the fact that homotopy equivalence is an equivalence relation, we obtain that M^b is homotopy-equivalent to M^a with a q-cell attached. □

Different stationarity concepts

We briefly review well-known definitions of various stationarity concepts and connections between them (see [1, 43, 44, 45, 46, 56]).

Definition 34. Let $\bar{x} \in M$.

(i) $\bar{x}$ is called weakly stationary if (4.2)–(4.5) hold and

$$\bar{\delta}_m^G \leq 0 \text{ for all } m \in I_{00}$$

(ii) $\bar{x}$ is called M-stationary if (4.2)–(4.5) hold and

$$\bar{\delta}_m^G \leq 0 \text{ and } \bar{\delta}_m^G \cdot \bar{\delta}_m^H = 0 \text{ for all } m \in I_{00}.$$

(iii) $\bar{x}$ is called strongly stationary if (4.2)–(4.5) hold and

$$\bar{\delta}_m^G = 0 \text{ and } \bar{\delta}_m^H \geq 0 \text{ for all } m \in I_{00}.$$

Note that a strongly stationary point is M-stationary and the latter is T-stationary. We see that M- and strongly stationary points describe local minima tighter than T-stationary points. Moreover, strong stationarity is the tightest condition for a local minimizer under the LICQ. It is worth mentioning that M-stationarity exhibits a full calculus in the sense of Mordukhovich (see [94]). The scheme in Figure 22 illustrates the stationarity concepts above.

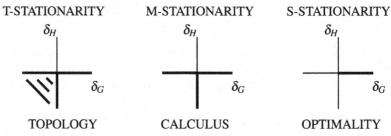

Figure 22 Stationarity concepts in MPVC

However, M- and strong stationarity exclude T-stationary points with $BI > 0$. These points are also crucial for the topological structure of the MPVC (see the cell-attachment theorem). For global optimization, points of T-index 1 play an important role. We emphasize that among the points of T-index 1 from a topological point of view there is no substantial difference between the points with $BI = 1$, $QI = 0$ and $BI = 0$, $QI = 1$. It is worth mentioning that a linear descent direction might exist in a nondegenerate T-stationary point with positive T-index. In particular, at points with $BI = 1$, $QI = 0$ there are exactly two directions of linear descent. Both of them are important from a global point of view. On the other hand, among weakly stationary points, there are those with negative and positive Lagrange multipliers corresponding to the same bi-active vanishing constraint. Due to the deformation theorem, such points are irrelevant for the topological structure of the MPVC.

We mention that the nondegeneracy assumption (as in Definition 30, ND4) cannot be stated for M- and strongly stationary points w.r.t. biactive vanishing constraints. This means that these points are singularities. Moreover, local minima for MPVC with bi-active vanishing constraints do not occur generically. We claim that their classification is sophisticated and might be established via singularity theory.

Links to MPCC

We point out that in Section 2.3 (see also [69]) the analogous stationarity concept for MPCCs turned out to be C-stationarity. Indeed, the MPCC feasible set can be described by nonsmooth equality constraints of minimum type. Moreover, generically the MPCC feasible set is a Lipschitz manifold of an appropriate dimension; that is, each nonsmooth equality constraint causes loss of one degree of freedom (see Section 2.2.2 and [70]). This permits the use of Clarke subdifferentials of these equality constraints to formulate the stationarity conditions, namely the C-stationarity. As C-stationarity is the topologically relevant stationarity concept for MPCCs, we consider it T-stationarity in the MPCC setting.

In contrast to the MPCC case, the MPVC feasible set (under the LICQ) is not a Lipschitz manifold but a set glued together from manifold pieces of **different dimensions** along their strata. Rather than by applying a general stationarity concept to MPVCs, like C-stationarity for MPCCs, T-stationarity for MPVCs is motivated by understanding the geometrical properties of a typical MPVC feasible set $\mathbb{V}$ directly, where $\mathbb{V}$ represents the solution set of the basic vanishing constraint relations $x \geq 0, xy \geq 0$.

A further analogy between C-stationarity for MPCCs and T-stationarity for MPVCs is established via convergence theory of certain regularization methods. In fact, the MPCC regularization method from [108] yields C-stationary points as limits of KKT points of the regularized problems ([108, Theorem 5.1]). The analogous limit points of an adaptation of this method to MPVCs from [47] are T-stationary.

Chapter 5
Bilevel Optimization

We study bilevel optimization problems from the optimistic perspective. The case of one-dimensional leader's variable is considered. Based on the classification of generalized critical points in one-parametric optimization, we describe the generic structure of the bilevel feasible set. Moreover, optimality conditions for bilevel problems are stated. In the case of higher-dimensional leader's variable, some bifurcation phenomena are discussed. The links to the singularity theory are elaborated.

5.1 Applications and examples

We consider bilevel optimization problems as hierarchical problems of two decision makers, the so-called leader and follower. The follower selects his decision knowing the choice of the leader, whereas the latter has to anticipate the follower's response in his decision. Bilevel programming problems have been studied in the monographs [5] and [15]. We model the bilevel optimization problem in the so-called optimistic formulation. To this aim, assume that the follower solves the parametric optimization problem (lower-level problem L)

$$L(x): \quad \min_y g(x,y) \quad \text{s.t.} \quad h_j(x,y) \geq 0, j \in J \tag{5.1}$$

and that the leader's optimization problem (upper-level problem U) is

$$U: \quad \min_{(x,y)} f(x,y) \quad \text{s.t.} \quad y \in \text{Argmin } L(x). \tag{5.2}$$

Above we have $x \in \mathbb{R}^n$, $y \in \mathbb{R}^m$, and the real-valued mappings $f, g, h_j, j \in J$ belong to $C^3(\mathbb{R}^n \times \mathbb{R}^m)$, $|J| < \infty$. Argmin $L(x)$ denotes the solution set of the optimization problem $L(x)$. For simplicity, additional (in)equality constraints in defining U are omitted.

In what follows, we present Stackelberg games as a classical application of bilevel programming. Furthermore, optimistic and pessimistic versions of bilevel optimization are compared. It turns out that the main difficulty in studying both versions lies in the fact that the lower level contains a global constraint. In fact, a point (x,y) is feasible if y solves a parametric optimization problem $L(x)$. It gives rise to studying the structure of the bilevel feasible set

$$M := \{(x,y)\,|\,y \in \text{Argmin } L(x)\}.$$

Finally, we give some guiding examples on the possible structure of M when $\dim(x) = 1$. We refer the reader to [75] for details.

Stackelberg game

In a Stackelberg game, there are two decision makers, the so-called leader and follower. The leader can adjust his variable $x \in \mathbb{R}^n$ at the upper level. This variable x as a parameter influences the follower's decision process $L(x)$ at the lower level. One might think of $L(x)$ being a minimization procedure w.r.t. the follower's variable $y \in \mathbb{R}^m$. Then, the follower chooses $y(x)$, a solution of $L(x)$ that is in general not unique. This solution $y(x)$ is anticipated by the leader at the upper level. After evaluating the leader's objective function $f(x,y(x))$, a new adjustment of x is performed. The game circles until the leader obtains an optimal parameter x w.r.t. his objective function f. The scheme in Figure 23 describes a Stackelberg game.

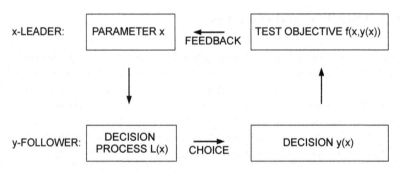

Figure 23 Stackelberg game

The difficulties in modeling such a two-hierarchy Stackelberg game come from its intrinsic dynamical behavior. In fact, the choice at the lower level is usually not unique. Hence, the feedback at the upper level cannot be prescribed a priori.

Pessimistic vs. Optimistic Versions

We model a Stackelberg game via bilevel optimization. For that, let the lower level be given by

$$L(x): \quad \min_y g(x,y) \quad \text{s.t.} \quad h_j(x,y) \geq 0, \, j \in J.$$

Using the follower's objective function f, we define

$$\varphi_o(x) := \min_y \{f(x,y) \mid y \in \text{Argmin } L(x)\}$$

and

$$\varphi_p(x) := \max_y \{f(x,y) \mid y \in \text{Argmin } L(x)\}.$$

where φ_o (resp. φ_p) is referred to an optimistic (resp. pessimistic) objective function value at the upper level. In the definition of φ_o, a best reply $y \in \text{Argmin } L(x)$ from the leader's point of view is assumed. It happens in the case of cooperation between the leader and the follower. Alternatively, if there is no cooperation between the players, the leader uses φ_p. In both situations, φ_o or φ_p are in general nonsmooth functions. To obtain a solution of a bilevel optimization problem, φ_o (resp. φ_p) is to be minimized w.r.t. x. The scheme in Figure 24 describes both optimistic and pessimistic perspectives.

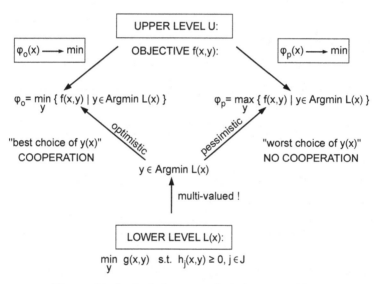

Figure 24 Optimistic vs. pessimistic perspective

The following example clarifies the difference between optimistic and pessimistic objective function values.

Example 26. Let $\dim(x) = 1$, $J = \emptyset$ and the graphs of $g(x, \cdot)$ be depicted in Figure 25 (for x close to $\bar{x}$). Clearly, $\text{Argmin } L(\bar{x}) = \{\bar{y}_1, \bar{y}_2\}$. For x close to $\bar{x}$, we obtain (see Figure 25)

$$\text{Argmin } L(x) = \begin{cases} \{y_1(x)\} & \text{if } x < \bar{x}, \\ \{y_2(x)\} & \text{if } x < \bar{x}. \end{cases}$$

Hence, we get with the leader's objective function f

$$\varphi_o(x) = \begin{cases} f(x, y_1(x)), & \text{if } x < \bar{x}, \\ \min\{f(\bar{x}, \bar{y}_1), f(\bar{x}, \bar{y}_1)\}, & \text{if } x = \bar{x}, \\ f(x, y_2(x)), & \text{if } x > \bar{x}, \end{cases}$$

$$\varphi_p(x) = \begin{cases} f(x, y_1(x)), & \text{if } x < \bar{x}, \\ \max\{f(\bar{x}, \bar{y}_1), f(\bar{x}, \bar{y}_1)\}, & \text{if } x = \bar{x}, \\ f(x, y_2(x)), & \text{if } x > \bar{x}. \end{cases}$$

We see that the only difference between $\varphi_o(x)$ and $\varphi_p(x)$ is the value at $\bar{x}$.

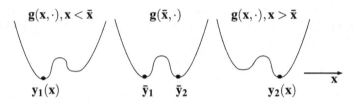

Figure 25 Graphs of $g(x, \cdot)$ from Example 26

Example 26 suggests that the main difficulty of bilevel programming lies in the structure of its feasible set M rather than in optimistic or pessimistic perspectives. Thus, we concentrate on the optimistic formulation in the subsequent analysis.

Examples

We present several typical examples for the case $\dim(x) = 1$. They motivate our results on the structure of the bilevel feasible set M. In all examples, the origin 0_{1+m} solves the bilevel problem U. Each example exhibits some kind of degeneracy in the lower level $L(x)$. Recall that $\dim(x) = 1$ throughout.

Example 27.

$$f(x,y) := -x + 2y_1 + \varphi(y_2, \ldots, y_m) \text{ with } \varphi \in C^3(\mathbb{R}^{m-1}, \mathbb{R}),$$
$$g(x,y) := (x - y_1)^2 + \sum_{j=2}^{m} y_j^2, J = \{1\} \text{ and } h_1(x,y) := y_1.$$

The degeneracy in the lower level $L(x)$ is the lack of strict complementarity at the origin 0_m.

The bilevel feasible set M becomes

$$M = \{(x, \max(x,0), 0, \ldots, 0) \mid x \in \mathbb{R}\}.$$

This example refers to Type 2 in the classification of Section 5.2.

Example 28.

$$f(x,y) := x + \sum_{j=1}^{m} y_j, g(x,y) := -y_1,$$

$$J = \{1\}, h_1(x,y) := x - \sum_{j=1}^{m} y_j^2.$$

The degeneracy in the lower level $L(x)$ is the violation of the so-called Mangasarian-Fromovitz constraint qualification (MFCQ) (see Section 5.2) at the origin 0_m. Moreover, the minimizer 0_m is a so-called Fritz John point but not a Karush-Kuhn-Tucker (KKT) point.

The bilevel feasible set M is a (half-)parabola,

$$M = \{(x, \sqrt{x}, 0, \ldots, 0) \mid x \geq 0\}.$$

This example refers to Type 4 in the classification of Section 5.2.

Example 29.

$$f(x,y) := x + \sum_{j=1}^{m} y_j, g(x,y) := \sum_{j=1}^{m} y_j, J = \{1, \ldots, m, m+1\},$$

$$h_j(x,y) := y_j, j = 1, \ldots, m, h_{m+1}(x,y) = x - \sum_{j=1}^{m} y_j.$$

The degeneracy in $L(0)$ is again the violation of the MFCQ at the origin 0_m. However, in contrast to Example 28, the minimizer 0_m is now a KKT point.

The bilevel feasible set M becomes

$$M = \{(x, 0, \ldots, 0) \mid x \geq 0\}.$$

This example refers to Type 5-1 in the classification of Section 5.2.

Example 30.

$$f(x,y) := -x + 2\sum_{j=1}^{m} y_j, g(x,y) := \sum_{j=1}^{m} jy_j, J = \{1, \ldots, m, m+1\},$$

$$h_j(x,y) := y_j, j = 1, \ldots, m, h_{m+1}(x,y) = -x + \sum_{j=1}^{m} y_j.$$

The degeneracy in $L(0)$ is the violation of the so-called linear independence constraint qualification (LICQ) at the origin 0_m, whereas the MFCQ is satisfied.

The bilevel feasible set M becomes

$$M = \{(x, \max(x, 0), 0, \ldots, 0) \mid x \in \mathbb{R}\}.$$

This example refers to Type 5-2 in the classification of Section 5.2.

Note that the feasible set M exhibits a kink in Examples 27 and 30, whereas it has a boundary in Examples 28 and 29. Moreover, the minimizer 0_m in $L(0)$ is strongly stable (in the terminology of Kojima [83]) in Examples 27 and 30 but not in Examples 28 and 29.

We note that, despite degeneracies in the lower level, the structure of the bilevel feasible set M with its kinks and boundaries remains stable under small C_s^3-perturbations of the defining functions.

5.2 Five types in parametric optimization

We consider the lower-level problem $L(\cdot)$ in a one-dimensional parametric optimization setting (i.e., $\dim(x) = 1$),

$$L(x): \quad \min_{y} g(x,y) \quad \text{s.t.} \quad h_j(x,y) \geq 0, \, j \in J.$$

We denote its feasible set by

$$M(x) := \left\{ y \in \mathbb{R}^m \mid h_j(x,y) \geq 0, \, j \in J \right\}$$

and for $\bar{y} \in M(\bar{x})$ the active index set by

$$J_0(\bar{x}, \bar{y}) := \left\{ j \in J \mid h_j(\bar{x}, \bar{y}) = 0 \right\}.$$

Definition 35 (Generalized critical point). A point $\bar{y} \in M(\bar{x})$ is called a generalized critical point (g.c. point) for $L(\bar{x})$ if the set of vectors

$$\left\{ D_y g(\bar{x}, \bar{y}), D_y h_j(\bar{x}, \bar{y}), \, j \in J_0(\bar{x}, \bar{y}) \right\} \tag{5.3}$$

is linearly dependent.

The critical set for $L(\cdot)$ is given by

$$\Sigma := \left\{ (x,y) \in \mathbb{R}^{1+m} \mid y \text{ is a g.c. point for } L(x) \right\}.$$

In [64], it is shown that generically each point of Σ is one of the Types 1–5. In what follows, we briefly discuss Types 1–5 and consider the structure of Σ locally around particular g.c. points that are local minimizers for $L(\cdot)$. Here, we focus on such parts of Σ that correspond to (local) minimizers,

$$\Sigma_{\min} := \{ (x,y) \in \Sigma \mid y \text{ is a local minimizer for } L(x) \},$$

in a neighborhood of $(\bar{x}, \bar{y}) \in \Sigma_{\min}$ (see [60]).

Points of Type 1

A point $(\bar{x}, \bar{y}) \in \Sigma$ is of Type 1 if $\bar{y}$ is a nondegenerate critical point for $L(\bar{x})$. This means that the following conditions ND1–ND3 hold.

ND1: The linear independence constraint qualification (LICQ) is satisfied at $(\bar{x}, \bar{y})$, meaning the set of vectors

$$\{D_y h_j(\bar{x}, \bar{y}),\ j \in J_0(\bar{x}, \bar{y})\} \tag{5.4}$$

is linearly independent.

From (5.3) and (5.4), we see that there exist (Lagrange multipliers) $\bar{\mu}_j,\ j \in J_0(\bar{x}, \bar{y})$, such that

$$D_y g(\bar{x}, \bar{y}) = \sum_{j \in J_0(\bar{x}, \bar{y})} \bar{\mu}_j D_y h_j(\bar{x}, \bar{y}). \tag{5.5}$$

ND2: $\bar{\mu}_j \neq 0,\ j \in J_0(\bar{x}, \bar{y})$.
ND3: $D^2_{yy} L(\bar{x}, \bar{y})_{|T_{\bar{y}}M(\bar{x})}$ is nonsingular.

Here, the matrix $D^2_{yy} L(\bar{x}, \bar{y})$ stands for the Hessian w.r.t. y variables of the Lagrange function L,

$$L(x, y) := g(x, y) - \sum_{j \in J_0(\bar{x}, \bar{y})} \bar{\mu}_j h_j(x, y), \tag{5.6}$$

and $T_{\bar{y}}M(\bar{x})$ denotes the tangent space of $M(\bar{x})$ at $\bar{y}$,

$$T_{\bar{y}}M(\bar{x}) := \{\xi \in \mathbb{R}^m \,|\, D_y h_j(\bar{x}, \bar{y}) \cdot \xi = 0,\ j \in J_0(\bar{x}, \bar{y})\}. \tag{5.7}$$

Condition ND3 means that the matrix $V^T D^2_{yy} L(\bar{x}, \bar{y}) V$ is nonsingular, where V is some matrix whose columns form a basis for the tangent space $T_{\bar{y}}M(\bar{x})$.

The linear index (LI) (resp. linear coindex, LCI) is defined to be the number of $\bar{\mu}_j$ in (5.5) that are negative (resp. positive). The quadratic index (QI) (resp. quadratic coindex, QCI) is defined to be the number of negative (resp. positive) eigenvalues of $D^2_{yy} L(\bar{x}, \bar{y})_{|T_{\bar{y}}M(\bar{x})}$.

Characteristic numbers: LI, LCI, QI, QCI

It is well-known that conditions ND1–ND3 allow us to apply the implicit function theorem and obtain unique C^2-mappings $y(x)$, $\mu_j(x)$, $j \in J_0(\bar{x}, \bar{y})$ in an open neighborhood of $\bar{x}$. It holds that $y(\bar{x}) = \bar{y}$ and $\mu_j(\bar{x}) = \bar{\mu}_j$, $j \in J_0(\bar{x}, \bar{y})$. Moreover, for x sufficiently close to $\bar{x}$, the point $y(x)$ is a nondegenerate critical point for $L(x)$ with Lagrange multipliers $\mu_j(x)$, $j \in J_0(\bar{x}, \bar{y})$ having the same indices LI, LCI, QI, QCI as $\bar{y}$. Hence, locally around $(\bar{x}, \bar{y})$ we can parameterize the set Σ by means of a unique C^2-map $x \mapsto (x, y(x))$. If $\bar{y}$ is additionally a local minimizer for $L(\bar{x})$ (i.e., LI $=$ QI $= 0$), then we get locally around $(\bar{x}, \bar{y})$

$$\Sigma_{\min} = \{(x, y(x)) \,|\, x \text{ sufficiently close to } \bar{x}\}.$$

Points of Type 2

A point $(\bar{x}, \bar{y}) \in \Sigma$ is of Type 2 if the following conditions A1–A6 hold:
  A1: The LICQ is satisfied at $(\bar{x}, \bar{y})$.
  A2: $J_0(\bar{x}, \bar{y}) \neq \emptyset$.
  After renumbering, we may assume that $J_0(\bar{x}, \bar{y}) = \{1, \ldots, p\}$, $p \geq 1$. Then, we have

$$D_y g(\bar{x}, \bar{y}) = \sum_{j=1}^{p} \bar{\mu}_j D_y h_j(\bar{x}, \bar{y}). \tag{5.8}$$

  A3: In (5.8), exactly one of the Lagrange multipliers vanishes.
  After renumbering, we may assume that $\bar{\mu}_p = 0$ and $\bar{\mu}_j \neq 0$, $j = 1, \ldots, p-1$.
  Let L and $T_{\bar{y}} M(\bar{x})$ be defined as in (5.6) and (5.7), respectively.
  A4: $D^2_{yy} L(\bar{x}, \bar{y})_{|T_{\bar{y}} M(\bar{x})}$ is nonsingular.
  We set

$$T_{\bar{y}}^+ M(\bar{x}) := \left\{ \xi \in \mathbb{R}^m \mid D_y h_j(\bar{x}, \bar{y}) \cdot \xi = 0, \ j \in J_0(\bar{x}, \bar{y}) \backslash \{p\} \right\}.$$

  A5: $D^2_{yy} L(\bar{x}, \bar{y})_{|T_{\bar{y}}^+ M(\bar{x})}$ is nonsingular.
  Let W be a matrix with m rows, whose columns form a basis of the linear space $T_{\bar{y}}^+ M(\bar{x})$. Put $\Phi = (h_1, \ldots, h_{p-1})^T$ and define the $m \times 1$-vectors

$$\alpha := -\left[\left(D_y \Phi \cdot D_y^T \Phi \right)^{-1} \cdot D_y \Phi \right]^T \cdot D_x \Phi,$$

$$\beta = -W \cdot \left(W^T \cdot D^2_{yy} L \cdot W \right)^{-1} \cdot W^T \left[D^2_{yy} L \cdot \alpha + D_x D_y^T L \right].$$

Note that all partial derivatives are evaluated at $(\bar{x}, \bar{y})$. Next, we put

$$\gamma := D_x h_p(\bar{x}, \bar{y}) + D_y h_p(\bar{x}, \bar{y})(\alpha + \beta).$$

  A6: $\gamma \neq 0$.
  Let δ_1 and δ_2 denote the number of negative eigenvalues of $D^2_{yy} L(\bar{x}, \bar{y})_{|T_{\bar{y}}^+ M(\bar{x})}$ and $D^2_{yy} L(\bar{x}, \bar{y})_{|T_{\bar{y}} M(\bar{x})}$, respectively, and put $\delta := \delta_1 - \delta_2$.

  Characteristic numbers: $\mathrm{sign}(\gamma), \ \delta$
  We proceed with the local analysis of the set Σ in a neighborhood of $(\bar{x}, \bar{y})$.
  (a) We consider the following associated optimization problem (without the p-th constraint):

$$\widetilde{L}(x): \ \underset{y \in \mathbb{R}^m}{\mathrm{minimize}} \ g(x, y) \ \text{ s.t. } \ h_j(x, y) \geq 0, \ j \in J \backslash \{p\}. \tag{5.9}$$

It is easy to see that $\bar{y}$ is a nondegenerate critical point for $\widetilde{L}(\bar{x})$ from A1, A3, and A5. As in Type 1, we get a unique C^2-map $x \mapsto (x, \widetilde{y}(x))$. This curve belongs to Σ as far as $\psi(x)$ is nonnegative, where

$$\psi(x) := h_p(x, \widetilde{y}(x)).$$

A few calculations show that

$$\frac{d\widetilde{y}(\bar{x})}{dx} = \alpha + \beta \text{ and hence } \frac{d\psi(\bar{x})}{dx} = \gamma. \tag{5.10}$$

Consequently, if we walk along the curve $x \mapsto (x, \widetilde{y}(x))$ as x increases, then at $x = \bar{x}$ we leave (or enter) the feasible set $M(x)$ according to $\text{sign}(\gamma) = -1(+1)$ (see A6).

(b) We consider the following associated optimization problem (with the p-th constraint as equality):

$$\widehat{L}(x): \quad \underset{y \in \mathbb{R}^m}{\text{minimize}} \; g(x, y) \;\; \text{s.t.} \;\; h_j(x, y) \geq 0, \; j \in J, \; h_p(x, y) = 0. \tag{5.11}$$

It is easy to see that $\bar{y}$ is a nondegenerate critical point for $\widehat{L}(\bar{x})$ from A1, A3, ad A4. Using results for Type 1, we get a unique C^2-map $x \mapsto (x, \widehat{y}(x))$. Note that $h_p(x, \widehat{y}(x)) \equiv 0$. Moreover, it can be calculated that

$$\text{sign}(\gamma) \cdot \text{sign}\left(\frac{d\mu_p(\bar{x})}{dx}\right) = -1 \; (\text{resp.} \; +1) \;\; \text{iff} \;\; \delta = 0 \; (\text{resp.} \; \delta = 1). \tag{5.12}$$

From this, since the curve $x \mapsto (x, \widetilde{y}(x))$ traverses the zero set "$h_p = 0$" at $(\bar{x}, \bar{y})$ transversally (see A6), it follows that $x \mapsto (x, \widetilde{y}(x))$ and $x \mapsto (x, \widehat{y}(x))$ intersect at $(\bar{x}, \bar{y})$ under a nonvanishing angle. Obviously, in a neighborhood of $(\bar{x}, \bar{y})$, the set Σ consists of $x \mapsto (x, \widehat{y}(x))$ and that part of $x \mapsto (x, \widetilde{y}(x))$ on which h_p is nonnegative.

Now additionally assume that $\bar{y}$ is a local minimizer for $L(\bar{x})$. Then, $\bar{\mu}_j > 0$, $j \in J_0(\bar{x}, \bar{y}) \backslash \{p\}$ in A3, and the matrix $D^2_{yy} L(\bar{x}, \bar{y})_{|T_{\bar{y}} M(\bar{x})}$ is positive definite in A4, and hence $\delta_2 = 0$.

We consider two cases for $\delta = 0$ or $\delta = 1$.

Case $\delta = 0$

In this case, $D^2_{yy} L(\bar{x}, \bar{y})_{|T^+_{\bar{y}} M(\bar{x})}$ is positive definite in A5. Hence, $\bar{y}$ is a strongly stable local minimizer for $L(\bar{x})$ (see [83] for details on the strong stability). Moreover, $\widetilde{y}(x)$ is a local minimizer for $L(x)$ if $h_p(x, \widetilde{y}(x)) > 0$. Otherwise, $\widehat{y}(x)$ is a local minimizer for $L(x)$ since the corresponding Lagrange multiplier $\mu_p(x)$ becomes positive due to (5.12). Note that the sign of $h_p(x, \widetilde{y}(x))$ corresponds to $\text{sign}(\gamma)$ as obtained in (5.10).

Then, we get locally around $(\bar{x}, \bar{y})$ that

$$\Sigma_{\min} = \left\{ (x, y(x)) \,|\, y(x) := \left\{ \begin{array}{l} \widetilde{y}(x), \; x \leq \bar{x} \\ \widehat{y}(x), \; \bar{x} \leq x \end{array} \right\} \right. \text{ if } \text{sign}(\gamma) = -1$$

and

$$\Sigma_{\min} = \left\{ (x, y(x)) \,|\, y(x) := \left\{ \begin{array}{l} \widehat{y}(x), \; x \leq \bar{x} \\ \widetilde{y}(x), \; \bar{x} \leq x \end{array} \right\} \right. \text{ if } \text{sign}(\gamma) = +1.$$

Case $\delta = 1$

In this case, $D^2_{yy}L(\bar{x},\bar{y})_{|T^+_{\bar{y}}M(\bar{x})}$ has exactly one negative eigenvalue. Thus, we obtain that the optimal value of the following optimization problem is negative:

$$\underset{\xi \in \mathbb{R}^m}{\text{minimize}} \, \xi^T \cdot D^2_{yy}g(\bar{x},\bar{y}) \cdot \xi \quad \text{s.t.} \quad \|\xi\| = 1, \, \xi \in T^+_{\bar{y}}M(\bar{x}), \, D_yh_p(\bar{x},\bar{y}) \cdot \xi \geq 0.$$

In view of that, at $(\bar{x},\bar{y})$ we can find a quadratic descent direction ξ for $L(\bar{x})$. Thus, $\bar{y}$ is not a local minimizer for $L(\bar{x})$, which contradicts the assumption above. We conclude that this case does not occur in $\Sigma_{\min}$.

Points of Type 3

A point $(\bar{x},\bar{y}) \in \Sigma$ is of Type 3 if the following conditions B1–B4 hold:

B1: LICQ is satisfied at $(\bar{x},\bar{y})$.

After renumbering, we may assume when $J_0(\bar{x},\bar{y}) \neq \emptyset$ that $J_0(\bar{x},\bar{y}) = \{1,\ldots,p\}$, $p \geq 1$. Then, we have

$$D_yg(\bar{x},\bar{y}) = \sum_{j=1}^{p} \bar{\mu}_j D_y h_j(\bar{x},\bar{y}). \tag{5.13}$$

B2: In (5.13), we have $\bar{\mu}_j \neq 0$, $j = 1,\ldots,p$.

Let L and $T_{\bar{y}}M(\bar{x})$ be defined as in (5.6) and (5.7), respectively.

B3: Exactly one eigenvalue of $D^2_{yy}L(\bar{x},\bar{y})_{|T_{\bar{y}}M(\bar{x})}$ vanishes.

Let V be a matrix whose columns form a basis for the tangent space $T_{\bar{y}}M(\bar{x})$. According to B3, let w be a nonvanishing vector such that $V^T \cdot D^2_{yy}L(\bar{x},\bar{y}) \cdot Vw = 0$, and put $v := V \cdot w$. Put $\Phi = (h_1,\ldots,h_{p-1})^T$, and define

$$\beta_1 := v^T(D^3_{yyy}L \cdot v)v - 3v^T D^2_{yy}L \cdot \left((D_y\Phi \cdot D_y^T\Phi)^{-1} \cdot D_y\Phi\right) \cdot (v^T D^2_{yy}\Phi v),$$

$$\beta_2 := D_x(D_yL \cdot v) - D_x^T\Phi \cdot \left((D_y\Phi \cdot D_y^T\Phi)^{-1} \cdot D_y\Phi\right) \cdot D^2_{yy}L \cdot v.$$

Note that all partial derivatives are evaluated at $(\bar{x},\bar{y})$. Next, we put

$$\beta := \beta_1 \cdot \beta_2.$$

B4: $\beta \neq 0$

Let α denote the number of negative eigenvalues of $D^2_{yy}L(\bar{x},\bar{y})_{|T_{\bar{y}}M(\bar{x})}$.

Characteristic numbers: sign(β), α

It turns out that in a neighborhood of $(\bar{x},\bar{y})$ the set Σ is a one-dimensional C^2-manifold. Moreover, the parameter x, viewed as a function on Σ, has a (nondegenerate) local maximum (resp. local minimum) at $(\bar{x},\bar{y})$ according to sign$(\beta) = +1$ (resp. sign$(\beta) = -1$). Consequently, the set Σ can be locally approximated by means of a parabola. In particular, if we approach the point $(\bar{x},\bar{y})$ along Σ, the path of local minimizers (with QI $= \alpha = 0$) stops and the local minimizer switches into a saddle point (with QI $= \alpha + 1 = 1$). Moreover, note that at $(\bar{x},\bar{y})$ there exists a unique (tangential) direction of cubic descent and hence $\bar{y}$ cannot be a local minimizer for $L(\bar{x})$. Hence, this case does not occur in $\Sigma_{\min}$.

Points of Type 4

A point $(\bar{x},\bar{y}) \in \Sigma$ is of Type 4 if the following conditions C1–C6 hold:

C1: $J_0(\bar{x},\bar{y}) \neq \emptyset$.

After renumbering, we may assume that $J_0(\bar{x},\bar{y}) = \{1,\ldots,p\}$, $p \geq 1$.

C2: $\dim \left\{ \mathrm{span} \left\{ D_y h_j(\bar{x},\bar{y}), \, j \in J_0(\bar{x},\bar{y}) \right\} \right\} = p - 1$.

C3: $p - 1 < m$.

From C2, we see that there exist $\bar{\mu}_j$, $j \in J_0(\bar{x},\bar{y})$, not all vanishing, such that

$$\sum_{j=1}^{p} \bar{\mu}_j D_y h_j(\bar{x},\bar{y}) = 0. \tag{5.14}$$

Note that the numbers $\bar{\mu}_j$, $j \in J_0(\bar{x},\bar{y})$ are unique up to a common multiple.

C4: $\bar{\mu}_j \neq 0$, $j \in J_0(\bar{x},\bar{y})$, and we normalize the $\bar{\mu}_j$'s by setting $\bar{\mu}_p = 1$.

Furthermore, we define

$$L(x,y) := h_p(x,y) + \sum_{j=1}^{p-1} \bar{\mu}_j h_j(x,y)$$

and

$$T_{\bar{y}} M(\bar{x}) := \left\{ \xi \in \mathbb{R}^m \, | \, D_y h_j(\bar{x},\bar{y}) \cdot \xi = 0, \, j \in J_0(\bar{x},\bar{y}) \right\}.$$

Let W be a matrix whose columns form a basis for $T_{\bar{y}} M(\bar{x})$. Define

$$A := D_x L \cdot W^T \cdot D_{yy}^2 L \cdot W \quad \text{and} \quad w := W^T \cdot D_y^T g,$$

all partial derivatives being evaluated at $(\bar{x},\bar{y})$.

C5: A is nonsingular.

Finally, define

$$\alpha := w^T \cdot A^{-1} \cdot w.$$

C6: $\alpha \neq 0$.

Let β denote the number of positive eigenvalues of A. Let γ be the number of negative $\bar{\mu}_j$, $j \in \{1,\ldots,p-1\}$, and put $\delta := D_x L(\bar{x},\bar{y})$.

Characteristic numbers: $\mathrm{sign}(\alpha)$, $\mathrm{sign}(\delta)$, γ, β

We proceed with the local analysis of the set Σ in a neighborhood of $(\bar{x},\bar{y})$. Conditions C2, C4, and C5 imply that (locally around $(\bar{x},\bar{y})$) at all points $(x,y) \in \Sigma$, apart from $(\bar{x},\bar{y})$, the LICQ holds. Moreover, the active set $J_0(\cdot)$ is locally constant $(= J_0(\bar{x},\bar{y}))$ on Σ. Having these facts in mind, we consider the map Ψ : $\mathbb{R} \times \mathbb{R}^m \times \mathbb{R}^{p-1} \times \mathbb{R} \longrightarrow \mathbb{R}^m \times R^p$:

$$\Psi(x,y,\mu,\lambda) := \begin{pmatrix} \lambda D_y g(x,y) + D_y h_p(x,y) + \sum_{j=1}^{p-1} \mu_j D_y h_j(x,y) \\ h_j(x,y) = 0, \ j = 1, \dots, p-1 \\ \lambda g(x,y) + h_p(x,t) + \sum_{j=1}^{p-1} \mu_j h_j(x,y) \end{pmatrix}.$$

Note that $\Psi(\bar{x},\bar{y},\bar{\mu},0) = 0$ and $D_{x,y,\mu}\Psi(\bar{x},\bar{y},\bar{\mu},0)$ is nonsingular due to C5 and C6. Hence, there exists the unique C^2-mapping $\lambda \mapsto (x(\lambda),y(\lambda),\mu(\lambda))$ such that $\Psi(x(\lambda),y(\lambda),\mu(\lambda),\lambda) \equiv 0$ and $(x(0),y(0),\mu(0)) = (\bar{x},\bar{y},\bar{\mu})$. Furthermore, it is not hard to see that locally around $(\bar{x},\bar{y})$

$$\Sigma = \{(x(\lambda),y(\lambda)) \,|\, \lambda \text{ sufficiently close to } 0\}.$$

The Lagrange multipliers corresponding to $(x(\lambda),y(\lambda))$ are

$$\left(-\frac{\mu_j(\lambda)}{\lambda}, \ j = 1, \dots, p-1, -\frac{1}{\lambda} \right). \tag{5.15}$$

It turns out that in a neighborhood of $(\bar{x},\bar{y})$ the set Σ is a one-dimensional C^2-manifold. The parameter x, viewed as a function on Σ, has a (nondegenerate) local maximum (resp. local minimum) at $(\bar{x},\bar{y})$ according to $\text{sign}(\alpha) = +1$ (resp. $\text{sign}(\alpha) = -1$). Consequently, the set Σ can be locally approximated by means of a parabola.

Now additionally assume that $\bar{y}$ is a local minimizer for $L(\bar{x})$. Then, $\bar{\mu}_j > 0$, $j = 1, \dots, p-1$ in C4, and hence

$$\gamma = 0. \tag{5.16}$$

Moreover, the matrix $W^T \cdot D_{yy}^2 L \cdot W$ is negative definite. In particular, we get

$$\beta = \begin{cases} n - (p-1) & \text{if } \text{sign}(\delta) = -1, \\ 0 & \text{if } \text{sign}(\delta) = 1. \end{cases} \tag{5.17}$$

We are interested in the local structure of $\Sigma_{\min}$ at $(\bar{x},\bar{y})$. It is clear from (5.15) that λ must be nonpositive if following the branch of local minimizers.

We consider two cases with respect to $\text{sign}(\alpha)$ and $\text{sign}(\delta)$.
Case 1: $\text{sign}(\alpha) = \text{sign}(\delta)$
A few calculations show that

$$D_\lambda g(x(\lambda),y(\lambda))_{\lambda=0} = -\alpha \cdot \delta.$$

Hence, $D_\lambda g(x(\lambda),y(\lambda))_{\lambda=0} < 0$ and $g(x(\cdot),y(\cdot))$ is strictly decreasing when passing $\lambda = 0$. Consequently, the possible branch of local minimizers corresponding to $\lambda \le 0$ cannot be one of global minimizers. We omit this case in view of our further interest in **global** minimizers in the context of bilevel programming problems.

Case 2: $\text{sign}(\alpha) \ne \text{sign}(\delta)$
In this case, we get locally around $(\bar{x},\bar{y})$ that

$$\Sigma_{\min} = \{(x(\lambda), y(\lambda)) \mid \lambda \leq 0\}.$$

In fact, for $\text{sign}(\alpha) = 1$ and $\text{sign}(\delta) = -1$, the linear and quadratic indices of $y(\lambda)$ for $L(x(\lambda))$, $\lambda < 0$, are

$$\text{LI} = \gamma = 0, \, QI = n - p - \beta + 1 = n - p - (n - p + 1) + 1 = 0.$$

For $\text{sign}(\alpha) = -1$ and $\text{sign}(\delta) = 1$, the linear and quadratic indices of $y(\lambda)$ for $L(x(\lambda))$, $\lambda < 0$, are

$$\text{LI} = \gamma = 0, \, QI = \beta = 0.$$

See (5.16) and (5.17) for the values of γ and β, respectively.

Points of Type 5

A point $(\bar{x}, \bar{y}) \in \Sigma$ is of Type 5 if the following conditions D1–D4 hold:
D1: $|J_0(\bar{x}, \bar{y})| = m + 1$.
D2: The set of vectors

$$\left\{ Dh_j(\bar{x}, \bar{y}), \, j \in J_0(\bar{x}, \bar{y}) \right\}$$

is linearly independent (derivatives in $\mathbb{R}^{m+1}$).

After renumbering we may assume that $J_0(\bar{x}, \bar{y}) = \{1, \ldots, p\}$, $p \geq 2$.

From D1 and D2, we see that there exist μ_j, $j \in J_0(\bar{x}, \bar{y})$, not all vanishing, such that

$$\sum_{j=1}^{p} \mu_j D_y h_j(\bar{x}, \bar{y}) = 0. \tag{5.18}$$

Note that the numbers μ_j, $j \in J_0(\bar{x}, \bar{y})$ are unique up to a common multiple.
D3: $\mu_j \neq 0$, $j \in J_0(\bar{x}, \bar{y})$.

From D1 and D2, it follows that there exist unique numbers β_j, $y \in J_0(\bar{x}, \bar{y})$ such that

$$Dg(\bar{x}, \bar{y}) = \sum_{j=1}^{p} \beta_j Dh_j(\bar{x}, \bar{y}). \tag{5.19}$$

Put

$$\Delta_{ij} := \beta_i - \beta_j \cdot \frac{\mu_i}{\mu_j} \text{ for } i, j = 1, \ldots, p,$$

and let Δ be the $p \times p$ matrix with Δ_{ij} as its (i, j)-th element.
D4: All off-diagonal elements of Δ do not vanish.
We set

$$L(\bar{x}, \bar{y}) = \sum_{j=1}^{p} \mu_j h_j(\bar{x}, \bar{y}).$$

From D2, we see that $D_x L(\bar{x}, \bar{y}) \neq 0$. We define

$$\gamma_j := \text{sign}\left(\mu_j \cdot D_x l(\bar{x}, \bar{y})\right) \text{ for } i, j = 1, \ldots, p.$$

By δ_j we denote the number of negative entries in the j-th column of Δ, $j = 1, \ldots, p$.

Characteristic numbers: γ_j, δ_j, $j = 1, \ldots, p$

We proceed with the local analysis of the set Σ in a neighborhood of $(\bar{x}, \bar{y})$. Conditions D1–D3 imply that locally around $(\bar{x}, \bar{y})$, at all points $(x, y) \in \Sigma \setminus \{(\bar{x}, \bar{y})\}$, the LICQ holds. Combining (5.18) and (5.19), we obtain

$$D_x g(\bar{x}, \bar{y}) = \sum_{j=1}^{p} \left(\beta_j - \beta_q \cdot \frac{\mu_j}{\mu_q} \right) D_x h_j(\bar{x}, \bar{y}), \quad q = 1, \ldots, p. \tag{5.20}$$

Both of these facts imply that for all $(x, y) \in \Sigma \setminus \{(\bar{x}, \bar{y})\}$ in a neighborhood of $(\bar{x}, \bar{y})$:

$$\|J_0(x, y)\| = m \text{ and } J_0(x, y) = J_0(\bar{x}, \bar{y}) \setminus \{q\} \tag{5.21}$$

with some $q \in \{1, \ldots, p\}$ (in general, depending on (x, y)). We put

$$M_q := \left\{ (x, y) \mid h_j(x, y) = 0, \ j \in J_0(\bar{x}, \bar{y}) \setminus \{q\} \right\}$$

and

$$M_q^+ := \left\{ (x, y) \in M_q \mid h_q(x, y) \geq 0 \right\}.$$

From (5.20) and (5.21), it is easy to see that locally around $(\bar{x}, \bar{y})$

$$\Sigma = \bigcup_{q=1}^{p} M_q^+.$$

The indices (LI, LCI, QI, QCI) along $M_q^+ \setminus \{(\bar{x}, \bar{y})\}$ are equal $(\delta_q, m - \delta_q, 0, 0)$. Let $q \in \{1, \ldots, p\}$ be fixed. M_q is a one-dimensional C^3-manifold from D2. Since the set of vectors

$$\left\{ D_y h_j(\bar{x}, \bar{y}), \ j \in J_0(\bar{x}, \bar{y}) \setminus \{q\} \right\}$$

is linearly independent, we can parameterize M_q by means of the unique C^3-mapping $x \mapsto (x, y^q(x))$ with $y^q(\bar{x}) = \bar{y}$. A short calculation shows that

$$\text{sign}\left(\frac{d h_q(x, y^q(x))}{dx} \bigg|_{x = \bar{x}} \right) = \gamma_q.$$

Hence, by increasing x, M_q^+ emanates from $(\bar{x}, \bar{y})$ (resp. ends at $(\bar{x}, \bar{y})$) according to $\gamma_q = +1$ (resp. $\gamma_q = -1$).

Now additionally assume that $\bar{y}$ is a local minimizer for $L(\bar{x})$. For describing $\Sigma_{\min}$ we define the so-called Karush-Kuhn-Tucker subset

$$\Sigma_{\text{KKT}} := \text{cl}\left\{ (x, y) \in \Sigma \mid (x, y) \text{ is of Type 1 with LI} = 0 \right\}.$$

It is shown in [64, Theorem 4.1] that, generically, Σ_{KKT} is a one-dimensional (piecewise C^2-) manifold with boundary. In particular, $(x,y) \in \Sigma_{KKT}$ is a boundary point iff at (x,y) we have $J_0(x,y) \neq \emptyset$ and the MFCQ fails to hold. We recall that the MFCQ is said to be satisfied for (x,y), $y \in M(x)$, if there exists a vector $\xi \in \mathbb{R}^m$ such that

$$D_y h_j(x,y) \cdot \xi > 0 \text{ for all } j \in J_0(x,y).$$

Now we consider two cases with respect to the signs of μ_j, $j \in J_0(\bar{x},\bar{y})$.

Case 1: all μ_j, $j \in J_0(\bar{x},\bar{y})$ have the same sign
Recalling (5.18), we obtain that MFCQ is not fulfilled at $(\bar{x},\bar{y})$. Hence, $(\bar{x},\bar{y})$ is a boundary point of Σ_{KKT}. Having in mind the formulas for the indices (LI $= \delta_q$, LCI $= m - \delta_q$, QI $= 0$, QCI $= 0$) along $M_q^+ \backslash \{(\bar{x},\bar{y})\}$, we obtain that $\delta_q = 0$ for some $q \in \{1,\dots,p\}$. Moreover, a simple calculation shows that

$$\Delta_{ij} = -\frac{\mu_i}{\mu_j} \cdot \Delta_{ji}, \; i, j = 1,\dots,p. \tag{5.22}$$

Since all μ_j, $j \in J_0(\bar{x},\bar{y})$, have the same sign, we get from (5.22) that

$$\text{sign}(\Delta_{ij}) = -\text{sign}(\Delta_{ji}), \; i, j = 1,\dots,p.$$

Hence,

$$\delta_j > 0 \text{ for all } j \in \{1,\dots,p\}\backslash\{q\}.$$

Finally, in this case we get locally around $(\bar{x},\bar{y})$ that

$$\Sigma_{\min} = \left\{ (x,y^q(x)) \,|\, x \geq \bar{x} \text{ (resp. } x \leq \bar{x}) \text{ if } \gamma_q = +1 \text{ (resp. } \gamma_q = -1), \delta_q = 0 \right\}.$$

We refer to this case as Type 5-1.

Case 2: μ_j, $j \in J_0(\bar{x},\bar{y})$ have different signs
The separation argument implies that the MFCQ is satisfied at $(\bar{x},\bar{y})$. Hence, a local minimizer $\bar{y}$ for $L(\bar{x})$ is also a KKT point, and $(\bar{x},\bar{y}) \in \Sigma_{KKT}$. From the MFCQ, $(\bar{x},\bar{y})$ is not a boundary point of Σ_{KKT}. Thus, there exist $q, r \in \{1,\dots,p\}, q \neq r$, such that

$$\delta_q = 0, \; \gamma_q = -1 \text{ and } \delta_r = 0, \; \gamma_r = +1.$$

Moreover, such q, r are unique fro (5.22), D4, and the definition of γ_j.
In this case, we get locally around $(\bar{x},\bar{y})$ that

$$\Sigma_{\min} = \left\{ (x,y(x)) \,|\, y(x) := \begin{cases} y^q(x), \; x \leq \bar{x} \text{ (if } \delta_q = 0, \gamma_q = -1) \\ y^r(x), \; x \geq \bar{x} \text{ (if } \delta_r = 0, \gamma_r = 1) \end{cases} \right\}.$$

We refer to this case as Type 5-2.

5.3 Structure of the feasible set: $\dim(x) = 1$

Our main goal is to describe the generic structure of the bilevel feasible set M, where

$$M := \{(x,y) \mid y \in \text{Argmin } L(x)\}.$$

The special case with unconstrained one-dimensional lower level (i.e., $J = \emptyset$ and $m = 1$) is treated in [17]. In that paper, the classification of one-dimensional singularities was heavily used, and for the higher-dimensional case (i.e., $m > 1$) it is conjectured that a similar result will hold.

However, the situation becomes extremely difficult to describe if inequality constraints are present in the lower level (i.e., $J \neq \emptyset$). In particular, kinks and ridges will appear in the feasible set, and such subsets might attract stable solutions of the bilevel problem. A simple example was presented in [17]. In this book, we restrict ourselves to the simplest case, where the x-dimension is equal to one (i.e., $n = 1$) but there are restrictions on the y-dimension. Then, the lower level $L(x)$ is a one-dimensional parametric optimization problem and we can exploit the well-known generic (five type) classification of so-called generalized critical points (see [64]) in order to describe the feasible set. Our main result (Theorems 41 and 42) states that, generically, the feasible set M is the union of C^2 curves with boundary points and kinks that can be parameterized by means of the variable x. The appearance of the boundary points and kinks is due to certain degeneracies of the corresponding local solutions in the lower level as well as the change from local to global solutions. Outside of the latter points, the feasible points $(x, y(x)) \in M$ correspond to nondegenerate minimizers of the lower level $L(x)$. Although $\dim(x) = 1$ might seem to be very restrictive, it should be noted that on typical curves in higher-dimensional x-space the one-dimensional features described here will reappear on those curves.

Obtaining the generic and stable structure of the feasible set M we, derive optimality criteria for the bilevel problem U. In order to guarantee the existence of solutions of the lower level, we will assume an appropriate compactness condition (5.24).

Simplicity of bilevel problems

First, we define simplicity of a bilevel programming problem at a feasible point. Recall again that $\dim(x) = 1$.

Definition 36 (Simplicity of Bilevel Problems). A bilevel programming problem U (with $\dim(x) = 1$) is called simple at $(\bar{x}, \bar{y}) \in M$ if one of the following cases occurs:

    Case I: Argmin $L(\bar{x}) = \{\bar{y}\}$ and $(\bar{x}, \bar{y})$ is of Type 1, 2, 4, 5-1, or 5-2,

    Case II: Argmin $L(\bar{x}) = \{\bar{y}_1, \bar{y}_2\}$ and $(\bar{x}, \bar{y}_1)$ and $(\bar{x}, \bar{y}_2)$ are both of Type 1, and
           additionally it holds that

$$\alpha := sign \left[\frac{d\left[g(x,y_2(x)) - g(x,y_1(x))\right]}{dx} \Bigg|_{x=\bar{x}} \right] \neq 0, \qquad (5.23)$$

where $y_1(x)$ and $y_2(x)$ are unique local minimizers for $L(x)$ in a neighborhood of $\bar{x}$ with $y_1(\bar{x}) = \bar{y}$ and $y_2(\bar{x}) = \bar{y}_2$ according to Type 1.

In order to avoid asymptotic effects, let $\mathcal{O}$ denote the set of $(g, h_j, j \in J) \in C^3(\mathbb{R}^{1+m}) \times \left[C^3(\mathbb{R}^{1+m})\right]^{|J|}$ such that

$$B_{g,h}(\bar{x}, c) \text{ is compact for all } (\bar{x}, c) \in \mathbb{R} \times \mathbb{R}, \qquad (5.24)$$

where

$$B_{g,h}(\bar{x}, c) := \{(x,y) \mid \|x - \bar{x}\| \leq 1, g(x,y) \leq c, y \in M(x)\}.$$

Note that $\mathcal{O}$ is C_s^3-open.

Now, we state our main result.

Theorem 41 (Simplicity is generic and stable). *Let $\mathcal{F}$ denote the set of defining functions $(f, g, h_j, j \in J) \in C^3(\mathbb{R}^{1+m}) \times \mathcal{O}$ such that the corresponding bilevel programming problem U is simple at all its feasible points $(\bar{x}, \bar{y}) \in M$. Then, $\mathcal{F}$ is C_s^3-open and C_s^3-dense in $C^3(\mathbb{R}^{1+m}) \times \mathcal{O}$.*

Proof. It is well-known from the one-dimensional parametric optimization ([64]) that generically the points of Σ are of Types 1–5 as defined above. Moreover, for the points of $M \subset \Sigma$, only Types 1, 2, 4, 5-1, or 5-2 may occur generically (see Section 5.2). Furthermore, the appearance of two different $y, z \in \text{Argmin } L(x)$ causes the loss of one degree of freedom from the equation

$$g(x, y) = g(x, z).$$

From the standard argument, by counting the dimension and codimension of the corresponding manifold in multijet space and by applying the multijet transversality theorem (see [63]), we get generically that

$$|\text{Argmin } L(x)| \leq 2.$$

Now, $|\text{Argmin } L(x)| = 1$ corresponds to Case I in Definition 36. For the case $|\text{Argmin } L(x)| = 2$, we obtain the points of Type 1. This comes from the fact that the appearance of Types 2, 4, 5-1, or 5-2 would cause another loss of freedom due to their degeneracy. Analogously, (5.23) in Case II is generically valid. The proof of the openness part is standard (see [63]). $\square$

Reduced bilevel feasible set

Using the description of $\Sigma_{\min}$ from Section 5.2, a reducible bilevel programming problem U can be locally reduced as follows.

Theorem 42 (Bilevel feasible set and reduced problem). *Let the bilevel programming problem U (with $\dim(x) = 1$) be simple at $(\bar{x}, \bar{y}) \in M$. Then, locally around $(\bar{x}, \bar{y})$, U is equivalent to the following reduced optimization problem:*

$$\text{Reduced Problem:} \quad \underset{(x,y) \in \mathbb{R}^1 \times \mathbb{R}^m}{\text{minimize}} \; f(x,y) \;\; s.t. \;\; (x,y) \in M_{loc}, \qquad (5.25)$$

where M_{loc} is given according to the cases in Definition 36:
Case I, Type 1:

$$M_{loc} = \{(x, y(x)) \,|\, x \text{ sufficiently close to } \bar{x}\},$$

Case I, Type 2:

$$M_{loc} = \left\{ (x, y(x)) \,|\, y(x) := \begin{cases} \widetilde{y}(x), & x \leq \bar{x} \\ \widehat{y}(x), & \bar{x} \leq x \end{cases} \right\} \; if \, sign(\gamma) = -1$$

or

$$M_{loc} = \left\{ (x, y(x)) \,|\, y(x) := \begin{cases} \widehat{y}(x), & x \leq \bar{x} \\ \widetilde{y}(x), & \bar{x} \leq x \end{cases} \right\} \; if \, sign(\gamma) = +1,$$

Case I, Type 4:

$$M_{loc} = \{(x(\lambda), y(\lambda)) \,|\, \lambda \leq 0\},$$

Case I, Type 5-1:

$$M_{loc} = \left\{ (x, y^q(x)) \,|\, x \geq \bar{x} \, (resp. \, x \leq \bar{x}) \; if \, \gamma_q = -1 \, (resp. \, \gamma_q = +1), \, \delta_q = 0 \right\},$$

Case I, Type 5-2:

$$M_{loc} = \left\{ (x, y(x)) \,|\, y(x) := \begin{cases} y^q(x), & x \leq \bar{x} \; (if \, \delta_q = 0, \, \gamma_q = -1) \\ y^r(x), & x \geq \bar{x} \; (if \, \delta_r = 0, \, \gamma_r = 1) \end{cases} \right\},$$

Case II:

$$M_{loc} = \{(x, y_1(x)) \,|\, x \geq \bar{x} \, (resp. \, x \leq \bar{x}) \; if \, \alpha = +1 \, (resp. \, \alpha = -1)\}.$$

We refer to Section 5.2 for details on Types 1, 2, 4, 5-1, and 5-2.

In each case, one of the possibilities for M_{loc} is depicted in Figure 26.

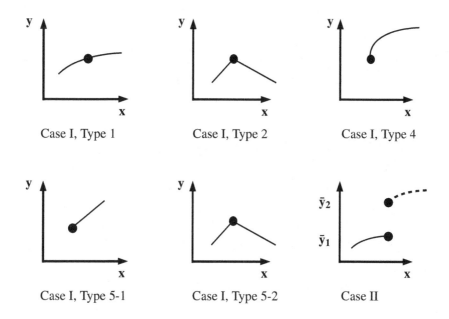

Case I, Type 1   Case I, Type 2   Case I, Type 4

Case I, Type 5-1   Case I, Type 5-2   Case II

Figure 26 Bilevel feasible set M_{loc} from Theorem 42

Optimality criteria for bilevel problems

Theorem 42 allows us to deduce optimality criteria for a reducible bilevel programming problem. In fact, the set M_{loc} from *Reduced Problem* is the feasible set of either a standard nonlinear optimization problem (NLP) (Case I, Types 1, 4, and 5-1 and Case II) or a mathematical programming problem with complementarity constraints (MPCC) (Case I, Types 2 and 5-2). Hence, we only need to use the corresponding optimality concepts of a Karush-Kuhn-Tucker point (for an NLP) and of an S-stationary point (for an MPCC); see [105] for the latter concept.

Theorem 43 (First-order optimality for simple bilevel problem). *Let a bilevel programming problem U (with $\dim(x) = 1$) be simple at its local minimizer $(\bar{x}, \bar{y}) \in M$. Then, according to the cases in Theorem 42, we obtain the following:*
Case I, Type 1:
$$D_x f(\bar{x}, \bar{y}) + D_y f(\bar{x}, \bar{y}) \cdot D_x y(\bar{x}) = 0,$$

Case I, Type 2:
$$[D_x f(\bar{x}, \bar{y}) + D_y f(\bar{x}, \bar{y}) \cdot D_x \widetilde{y}(\bar{x})] \leq 0,$$
$$[D_x f(\bar{x}, \bar{y}) + D_y f(\bar{x}, \bar{y}) \cdot D_x \widehat{y}(\bar{x})] \leq 0,$$

if $sign(\gamma) = -1$ or

$$[D_x f(\bar{x}, \bar{y}) + D_y f(\bar{x}, \bar{y}) \cdot D_x \widehat{y}(\bar{x})] \leq 0,$$

$$[D_x f(\bar{x}, \bar{y}) + D_y f(\bar{x}, \bar{y}) \cdot D_x \tilde{y}(\bar{x})] \le 0,$$

if sign(γ) = +1,
Case I, Type 4:

$$D_x f(\bar{x}, \bar{y}) \cdot D_\lambda x(0) + D_y f(\bar{x}, \bar{y}) \cdot D_\lambda y(0) \le 0,$$

Case I, Type 5-1:

$$D_x f(\bar{x}, \bar{y}) + D_y f(\bar{x}, \bar{y}) \cdot D_x y^q(\bar{x}) \ge 0, \; if \; \gamma_q = -1, \; \delta_q = 0$$

or

$$D_x f(\bar{x}, \bar{y}) + D_y f(\bar{x}, \bar{y}) \cdot D_x y^q(\bar{x}) \le 0, \; if \; \gamma_q = +1, \; \delta_q = 0,$$

Case I, Type 5-2:

$$[D_x f(\bar{x}, \bar{y}) + D_y f(\bar{x}, \bar{y}) \cdot D_x y^q(\bar{x})] \le 0,$$

$$[D_x f(\bar{x}, \bar{y}) + D_y f(\bar{x}, \bar{y}) \cdot D_x y^r(\bar{x})] \le 0,$$

Case II:

$$D_x f(\bar{x}, \bar{y}) + D_y f(\bar{x}, \bar{y}) \cdot D_x y(\bar{x}) \ge 0, \; if \; \alpha = -1,$$

or

$$D_x f(\bar{x}, \bar{y}) + D_y f(\bar{x}, \bar{y}) \cdot D_x y(\bar{x}) \le 0, \; if \; \alpha = +1.$$

Note that the derivatives of implicit functions above can be obtained from the defining equations as discussed in Section 5.2.

5.4 Toward the case $dim(x) \ge 2$

In the higher-dimensional case (i.e., $dim(x) \ge 2$) there will appear more complicated singularities in the description of the feasible set. In particular, we will present stable examples when more than one Lagrange multiplier vanishes. This will be an extension of Type 2 (see Examples 31 and 32). Here, combinatorial and bifurcation issues occur. On the other hand, we will not be able to describe all generic situations. This obstruction comes from classification in singularity theory. In fact, in one variable (y) there is already a countable infinite list of local minimizers: in the unconstrained case, the functions y^{2k}, $k \ge 1$ and in the constrained case $y \ge 0$ the functions y^k, $k \ge 1$. However, a complete list of local minimizers for functions of two variables or more is not even known. Therefore, we have to bring the objective function of the bilevel problem into play as well. If we restrict ourselves to a neighborhood of a (local) solution of the bilevel problem, then the generic situation becomes easier. For example, the above-mentioned singularities y^{2k} ($k \ge 2$) as well as the constrained singularities y^k ($k \ge 3$), $y \ge 0$, can generically be avoided in local solutions of the bilevel problem. The key idea is explained below and illustrated in Examples 33 and 34.

Combinatorial and bifurcation issues

Remark 30. We note that all singularities appearing for lower-dimensional x may reappear at higher-dimensional x in a kind of product structure. In fact, the lower-dimensional singularity may appear as a normal section in the corresponding normal-tangential stratification (see [29]). For example, let $x = (x_1, x_2, \ldots, x_n)$ and let the lower-level problem $L(x)$ be

$$L(x) : \quad \min_y \ (y - x_1)^2 \text{ s.t. } y \geq 0.$$

Then, the feasible set M becomes

$$M = \{(x_1, x_2, \ldots, x_n, \max\{x_1, 0\} \,|\, x \in \mathbb{R}^n\},$$

and in this particular case we see that M is diffeomorphic to product,

$$\{(x_1, \max\{x_1, 0\}) \,|\, x_1 \in \mathbb{R}\} \times \mathbb{R}^{n-1}.$$

At this point, we come to typical examples with several vanishing Lagrange multipliers. Here, we assume that the LICQ at the lower level is fulfilled, that the dimensions of the variables x and y coincide (i.e., $n = m$), that $J_0(\bar{x}, \bar{y}) = m$, and that $\bar{x} = \bar{y} = 0$. Taking the constraints h_j as new coordinates, we may assume that the lower-level feasible set $M(0)$ is just the nonnegative orthant. In this setting, the Lagrange multipliers of the lower-level function g at the origin just become the partial derivatives with respect to the coordinates y_j, $j = 1, \ldots, m$. Now we suppose that all these partial derivatives vanish (generalization of Type 2). Then, the Hessian $D^2_{yy}g(0,0)$ comes into play and we assume that it is nonsingular. In order that the origin be a (local) minimizer for $L(0)$, a stable condition becomes that the positive cone of the Hessian $D^2_{yy}g(0,0)$ contains the nonnegative orthant with deleted origin. This gives rise to several combinatorial possibilities, depending on the number of negative eigenvalues of $D^2_{yy}g(0,0)$. In the next two examples, we restrict ourselves to two dimensions, $n = m = 2$.

Example 31. In this example the Hessian $D^2_{yy}g(0,0)$ has two (typically distinct) positive eigenvalues. In particular, $D^2_{yy}g(0,0)$ is positive definite:

$$f(x_1, x_2, y_1, y_2) = (-x_1 + 2y_1) + (-x_2 + 2y_2),$$

$$L(x_1, x_2) : \quad \min_y \ g(x_1, x_2, y_1, y_2) := (y_1 - x_1)^2 + (y_1 - x_1) \cdot (y_2 - x_2) + (y_2 - x_2)^2$$

$$\text{s.t. } y_1 \geq 0, y_2 \geq 0.$$

In order to obtain the feasible set M, we have to consider critical points of $L(x_1, x_2)$ for the following four cases I–IV. These cases result from the natural stratification of the nonnegative orthant in y-space:

$$I ; y_1 > 0, y_2 > 0 \quad II : y_1 = 0, y_2 > 0$$
$$III : y_1 > 0, y_2 = 0 \quad IV : y_1 = 0, y_2 = 0.$$

It turns out that the feasible set M is a piecewise smooth two-dimensional manifold. Moreover, it can be parameterized via the x-variable by means of a subdivision of the x-space into four regions according to Cases I–IV above (see Figure 27).

On the regions I–IV, the corresponding global minimizer $(y_1(\cdot), y_2(\cdot))$ is given by

$$(y_1(x), y_2(x)) = \begin{cases} (x_1, x_2), & \text{if } (x_1, x_2) \in I, \\ (0, \frac{x_1}{2} + x_2), & \text{if } (x_1, x_2) \in II, \\ (\frac{x_2}{2} + x_1, 0), & \text{if } (x_1, x_2) \in III, \\ (0, 0), & \text{if } (x_1, x_2) \in IV. \end{cases} \qquad (5.26)$$

In particular, we obtain $M = \{(x, y(x)) \mid y(x) \text{ as in } (5.26)\}$. A few calculations show that the origin $(0, 0)$ solves the corresponding bilevel problem U.

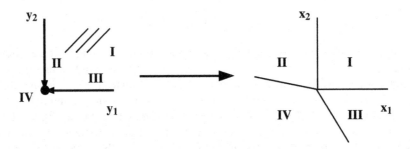

Figure 27 Illustration of Example 31

Example 32. In this example, the Hessian $D_{yy}^2 g(0, 0)$ has one positive and one negative eigenvalue:

$$f(x_1, x_2, y_1, y_2) = -3x_1 + x_2 + 4y_1 + 5y_2,$$

$$L(x_1, x_2) : \quad \min_y g(x_1, x_2, y_1, y_2) := (y_1 - x_1)^2 + 4(y_1 - x_1) \cdot y_2 + 3\left(y_2 + \frac{1}{3}x_2\right)^2$$

$$\text{s.t. } y_1 \geq 0, y_2 \geq 0.$$

It is easy to see that $(y_1, y_2) = (0, 0)$ is the global minimizer for $L(0, 0)$. Analogously to Example 31, we subdivide the parameter space (x_1, x_2) into regions on which the global minimizer $(y_1(x), y_2(x))$ for L(x) is a smooth function. Here, we obtain three regions, II–IV (see Figure 28). Note that the region corresponding to Case I is empty.

In addition, for the parameters (x_1, x_2) lying on the half-line

$$G : x_1 = (2 + \sqrt{3})x_2, x_1 \geq 0,$$

the problem $L(x)$ exhibits two different global minimizers. This is due to the fact that $(y_1, y_2) = (0, 0)$ is a saddle point of the objective function $g(0, y_1, y_2)$. Moreover, $(y_1, y_2) = (0, 0)$ is not strongly stable for $L(0, 0)$.

On the regions II–IV and G, the corresponding global minimizers $(y_1(\cdot), y_2(\cdot))$ are given by

$$(y_1(x), y_2(x)) = \begin{cases} (0, \frac{2}{3}x_1 - \frac{1}{3}x_2), & \text{if } (x_1, x_2) \in II, \\ (x_1, 0), & \text{if } (x_1, x_2) \in III, \\ (0, 0), & \text{if } (x_1, x_2) \in IV, \\ \{(0, \frac{2}{3}x_1 - \frac{1}{3}x_2), (x_1, 0)\} & \text{if } (x_1, x_2) \in G. \end{cases} \tag{5.27}$$

Here, $M = \{(x, y(x)) \mid y(x) \text{ as in } (5.27)\}$. We point out that the bilevel feasible set M is now a two-dimensional nonsmooth Lipschitz manifold with boundary, but it cannot be parameterized by the variable x. Again, one calculates that the origin $(0, 0)$ solves the corresponding bilevel problem U.

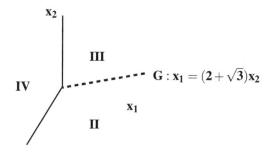

Figure 28 Illustration of Example 32

Remark 31. Let us consider in Examples 31 and 32 a smooth curve around the origin that traverses the partition of x-space in a transverse way, for example a circle C. Then, restricted to C, the dimension of x reduces to 1 and we rediscover a simple bilevel problem.

Avoiding higher-order singularities

In order to avoid certain higher-order singularities in the description of the feasible set M, we have to focus on a neighborhood of (local) solutions of the bilevel problem. The key idea is as follows. Suppose that the feasible set M contains a smooth curve, say C, through the point $(\bar{x}, \bar{y}) \in M$. Let the point $(\bar{x}, \bar{y})$ be a local solution of the bilevel problem U (i.e., $(\bar{x}, \bar{y})$ is a local minimizer for the objective function f on the set M). Then, $(\bar{x}, \bar{y})$ is also a local minimizer for f restricted to the curve C. If, in addition, $(\bar{x}, \bar{y})$ is a nondegenerate local minimizer for $f_{|C}$, then we may shift this local minimizer along C by means of a linear perturbation of f. After that perturbation with resulting $\tilde{f}$, the point $(\bar{x}, \bar{y})$ is no longer a local minimizer for $\tilde{f}_{|C}$ and hence no longer a local minimizer for $\tilde{f}_{|M}$. Now, if the singularities in M outside of the point $(\bar{x}, \bar{y})$ are of lower order, then we are able to move away from the higher-order singularity. This simple idea was used in particular in [17]. The key point, however, is to find a smooth curve through a given point of the feasible set *M*. An illustration will be presented in Examples 33 and 34. In contrast, note that in Examples 31 and 32 such a smooth curve through the origin $(0,0)$ does not exist.

Example 33. Consider the one-dimensional functions y^{2k}, $k = 1, 2, \ldots$. The origin $y = 0$ is always the global minimizer. For $k = 1$, it is nondegenerate (Type 1), but for $k \geq 2$ it is degenerate. Let $k \geq 2$ and $x = (x_1, x_2, \ldots, x_{2k-2})$. Then the function $g(x,y)$ with x as parameter,

$$g(x,y) = y^{2k} + x_{2k-2}y^{2k-2} + x_{2k-3}y^{2k-3} + \ldots + x_1 y,$$

is a so-called universal unfolding of the singularity y^{2k}. Moreover, the singularities with respect to *y* have a lower codimension (i.e., lower order) outside the origin $x = 0$ (see [3, 11]). Consider the unconstrained lower-level problem

$$L(x): \quad \min_{y} g(x,y)$$

with corresponding bilevel feasible set *M*. Let the smooth curve *C* in (x,y)-space be defined by the equations

$$x_1 = x_2 = \ldots = x_{2k-3} = 0, \; ky^2 + (k-1)x_{2k-2} = 0.$$

It is not difficult to see that, indeed, *C* contains the origin and belongs to the bilevel feasible set *M*.

Example 34. Consider the one-dimensional functions y^k, $k \geq 1$ under the constraint $y \geq 0$. The origin $y = 0$ is always the global minimizer. The case $k = 1$ is nondegenerate (Type 1), whereas the case $k = 2$ corresponds to Type 2. Let $k \geq 3$ and $x = (x_1, x_2, \ldots, x_{k-1})$. Then, analogously to Example 33, the function $g(x,y)$,

$$g(x,y) = y^k + x_{k-1}y^{k-1} + x_{k-2}y^{k-2} + \ldots + x_1 y, \; y \geq 0,$$

is the universal unfolding of the (constrained) singularity y^k, $y \geq 0$. Consider the constrained lower-level problem

$$L(x): \quad \min_{y} g(x,y) \text{ s.t. } y \geq 0$$

with corresponding bilevel feasible set *M*.

In order to find a smooth curve *C* through the origin and belonging to *M*, we put

$$x_1 = x_2 = \ldots = x_{k-3} = 0.$$

So, we are left with the reduced lower-level problem function

$$\widetilde{L}(x_{k-2}, x_{k-1}): \quad \min_{y} \widetilde{g}(x_{k-2}, x_{k-1}, y) \text{ s. t. } y \geq 0$$

with reduced feasible set $\widetilde{M}$, where

$$\widetilde{g}(x_{k-2}, x_{k-1}, y) = y^k + x_{k-1} y^{k-1} + x_{k-2} y^{k-2}.$$

First, let $x_{k-1} < 0$ and $x_{k-2} > 0$, and consider the curve defined by the equation

$$x_{k-2} - \frac{1}{4} x_{k-1}^2 = 0.$$

One calculates that, for points on this curve, the lower level $\widetilde{L}$ has two different global minimizers on the set $y \geq 0$ (with $\widetilde{g}$-value zero), one of them being $y = 0$. Second, we note that the set $\{(x_{k-1}, x_{k-2}, 0) \mid x_{k-1} \geq 0, x_{k-2} \geq 0\}$ belongs to $\widetilde{M}$. From this, we obtain that the curve C defined by the equations

$$x_1 = x_2 = \ldots = x_{k-3} = y = 0, \ x_{k-2} - \frac{1}{4} x_{k-1}^2 = 0,$$

belongs to M.

Finally, we remark that a complete, systematic generic description of the feasible bilevel set M in a neighborhood of local solutions of the bilevel problem U for higher x-dimensions is a very challenging issue for future research. Another interesting point for future research would be the discovery of a stable generic constraint qualification under which the whole feasible set M might be expected to be a Lipschitz manifold with boundary.

Chapter 6
Impacts on Nonsmooth Analysis

We discuss the notions of regular and critical points/values for nonsmooth functions. The notion of topologically regular points for min-type functions is introduced. It is shown that the level set of a min-type function corresponding to a regular value, is a Lipschitz manifold. The application of the Clarke's implicit function theorem is discussed here. A nonsmooth version of Sard's Theorem for min-type functions is shown. Finally, we discuss the links to metrically regular and critical points/values of nonsmooth functions.

6.1 Criticality for nonsmooth functions

The crucial notion of analysis is that of regular/critical points for a function $F : \mathbb{R}^n \longrightarrow \mathbb{R}^k$. It is well-known that in the case of a smooth F the surjectivity of its Jacobian $DF(\bar{x})$ provides regularity at $\bar{x}$. In the nonsmooth case, we do have a discrepancy between the different suggested concepts. Moreover, it turns out that we have to take into account different classes of nonsmooth functions. Two main questions should be addressed when developing nonsmooth analysis:

(1)   What kind of nonsmooth functions do we study?
(2)   How can one define regular/critical points?

In the following scheme (see Figure 29), we illustrate what cases (regarding the questions above) we deal with.

Nonsmooth Functions	Regular/Critical Points
Tame functions $\mathbb{R}^n \longrightarrow \mathbb{R}^k$ in o-minimal structures	Metric regularity
Lipschitz functions $\mathbb{R}^n \longrightarrow \mathbb{R}$ with Whitney-stratifiable graphs	via Clarke's subdifferential
Min-type functions $\mathbb{R}^n \longrightarrow \mathbb{R}^k$	via MFC

Figure 29 Nonsmooth analysis

We point out that the considerations of tame functions and metric regularity rely on [50], and those of Lipschitz functions $\mathbb{R}^n \longrightarrow \mathbb{R}$ and Clarke's subdifferentials rely on [10].

Clearly, question (2) means that some desirable properties should hold at regular points, such as metric or topological properties. Section 6.3 will be devoted to this issue. Furthermore, the set of critical values needs to be of Lebesgue measure zero. This is a version of the classical Sard's Theorem and is the focus of Section 6.2.

6.2 Versions of Sard's Theorem

In this section, we first recall the classical Sard's Theorem. Furthermore, nonsmooth versions of Sard's Theorem are provided for

(i)   tame functions via a metric regularity notion (see [50]),
(ii)   stratifiable functions via Clarke's subdifferentials (see [10]),
(iii)   min-type functions via the Mangasarian-Fromovitz condition (see Section 2.2.1).

Smooth case

Definition 37 (Regular/critical points for smooth F; see, e.g., [63]). Let $F \in C^\infty(\mathbb{R}^n, \mathbb{R}^k)$. A point $\bar{x} \in \mathbb{R}^n$ is called critical if the linear map from $\mathbb{R}^n$ to $\mathbb{R}^k$ given by $\xi \mapsto DF(\bar{x})\xi$ is not surjective. In other words, $\bar{x}$ is regular (i.e., not critical) if and only if $DF(\bar{x})[\mathbb{R}^n] = \mathbb{R}^k$. A point $y \in \mathbb{R}^k$ is called a regular (resp. critical) value for F if $F^{-1}(y)$ contains no critical points (resp. contains at least one critical point).

Remark 32. In the following cases, the criticality of $\bar{x} \in \mathbb{R}^n$ is equivalent to

(a) $k = 1 : DF(\bar{x}) = 0$,
(b) $k > n$: every $\bar{x} \in \mathbb{R}^n$ is critical, and
(c) $1 \leq k \leq n$: $D^T F_i(\bar{x})$, $i = 1, \ldots, k$ are linearly independent.

The following theorem is well-known.

Theorem 44 (Sard's Theorem; see, e.g., [63]). *The set of critical values of $F \in C^\infty(\mathbb{R}^n, \mathbb{R}^k)$ has Lebesgue measure zero.*

We point out that Sard's Theorem remains true for $F \in C^m(\mathbb{R}^n, \mathbb{R}^k)$, provided that $m > \max(n - k, 0)$ (see [111, 117]). Thus, a function $F \in C^1(\mathbb{R}^n, \mathbb{R}^k)$, $k > n$, can never be surjective. However, there exist surjective continuous functions from $\mathbb{R}^n$ to $\mathbb{R}^k$, $k > n$ (see the Peano space-filling curve from [107]).

Tame functions and metric regularity

We recall the notions of o-minimal structures and corresponding tame functions (see [14, 21, 50]).

Definition 38 (o-minimal structure; see, e.g., [14, 21]). A structure on $\mathbb{R}$ is a sequence $\mathscr{S} = (\mathscr{S}_n)$, $n \in \mathbb{N}$ such that

(D1) $\mathscr{S}_n$ is a Boolean algebra of subsets of $\mathbb{R}^n$ (i.e., $\emptyset \in S_n$ and $\mathscr{S}_n$ contains unions, intersections, and complements of its elements),
(D2) if $A \in \mathscr{S}_n$, then $A \times \mathbb{R}, \mathbb{R} \times A \in \mathscr{S}_{n+1}$,
(D3) $\{(x_1, \ldots, x_n) \,|\, x_i = x_j\} \in \mathscr{S}_n$ for all $1 \le i < j \le n$,
(D4) if $A \in \mathscr{S}_{n+1}$, then $\pi(A) \in \mathscr{S}_n$, where $\pi : (x, x_{n+1}) \mapsto x$ is the projection onto $\mathbb{R}^n$.

A structure is called o-minimal if in addition

(D5) $\{(x, y) \in \mathbb{R}^2 \,|\, x < y\} \in \mathscr{S}_2$ and
(D6) the elements of $\mathscr{S}_1$ are finite unions of points and open intervals.

The elements of $\mathscr{S}_n$, $n \in \mathbb{N}$ are called definable in $\mathscr{S}$.

We give some examples of o-minimal structures.

Example 35 (Examples of o-minimal structures; see, e.g., [14, 21]).

(i) **Semialgebraic sets** are finite unions of the sets

$$\{x \in R^n \,|\, p_i(x) < 0, i \in I, q_j(x) = 0, j \in J\},$$

where I and J are finite index sets and p_i, $i \in I$ and q_j, $j \in J$ are polynomials. Note that the validity of $(D4)$ is due to the nontrivial Tarski-Seidenberg Theorem (see [8]).

(ii) **Globally subanalytic sets:** As above, semianalytic sets can be constructed as finite unions of the sets

$$\{x \in R^n \,|\, f_i(x) < 0, i \in I, g_j(x) = 0, j \in J\},$$

where I and J are finite index sets and f_i, $i \in I$ and g_j, $j \in J$ are real analytic functions. A set $A \subset \mathbb{R}^n$ is called subanalytic if for any $x \in A$ there is an open neighborhood U of x and a bounded semianalytic set $S \subset \mathbb{R}^{n+m}$ such that the projection of S onto $\mathbb{R}^n$ is $A \cap U$. Finally, a set $B \subset \mathbb{R}^n$ is called globally subanalytic if $G(B)$ is subanalytic, where G is a semialgebraic homeomorphism of $\mathbb{R}^n$ onto $(-1, 1)^n$.

Definition 39 (Tame sets and functions; see, e.g., [14, 21]). Let $\mathscr{S}$ be an o-minimal structure on $\mathbb{R}$. A set A is called tame if its intersection with any bounded definable set is definable in $\mathscr{S}$. A function $F : \mathbb{R}^n \longrightarrow \mathbb{R}^k$ is definable (tame) if its graph is a definable (tame) set in $\mathscr{S}$.

Remark 33. Note that definable and tame sets are closed under interior and closure operations. Definable and tame functions are closed under infimum and supremum.

We recall the notion of metrically regular/critical points.

Definition 40 (Metrically regular/critical points [49]). A point $\bar{x}$ is called metrically regular for $F : \mathbb{R}^n \longrightarrow \mathbb{R}^k$ if there exist $L > 0$ and neighborhoods U and V of $\bar{x}$ and $F(\bar{x})$, respectively such that

$$\text{dist}(x, F^{-1}(y)) \leq L \text{dist}(y, F(x)) \text{ for all } x \in U, y \in V.$$

Otherwise, $\bar{x}$ is called metrically critical. A point $y \in \mathbb{R}^k$ is called a metrically regular (resp. critical) value for F if $F^{-1}(y)$ contains no metrically critical points (resp. contains at least one metrically critical point).

Now we are ready to state the nonsmooth version of Sard's Theorem for tame functions.

Theorem 45 (Sard's Theorem for tame functions [50]). *The set of metrically critical values of a tame function $F : \mathbb{R}^n \longrightarrow \mathbb{R}^k$ has Lebesgue measure zero.*

The proof of Theorem 45 is mainly based on results from tame geometry, such as monotonicity theorem and cell-decomposition theorem (see. [50]).

Stratifiable functions and Clarke's subdifferentials

Here, we consider Lipschitz functions $f : \mathbb{R}^n \longrightarrow \mathbb{R}$ whose graphs admit a C^∞-Whitney stratification.

Definition 41 (Whitney stratification [91]). A C^∞-stratification $\mathscr{X} = (X_i)$, $i \in \mathbb{N}$ of $X \subset \mathbb{R}^n$ is a locally finite partition of X into C^∞ manifolds $X_i \subset \mathbb{R}^n$ (called strata of X) such that

$$\text{if } \overline{X_i} \cap X_j \neq 0, \text{ then } X_j \subset \overline{X_i} \backslash X_i.$$

A C^∞-stratification $\mathscr{X} = (X_i)$, $i \in \mathbb{N}$ of $X \subset \mathbb{R}^n$ is called Whitney stratification if for each $x \in \overline{X_i} \cap X_j$, $(i \neq j)$ and for each sequence $(x_l) \subset X_i$ it holds that

$$\text{if } x_l \longrightarrow x, T_{x_l} X_i \longrightarrow \mathscr{T}, \text{ then } T_x X_j \subset \mathscr{T}.$$

Here, $T_{x_l} X_i$ denotes the tangent space of X_i at x_l.

Definition 42 (Whitney stratifiable functions). A function $f : \mathbb{R}^n \longrightarrow \mathbb{R}$ is called C^∞-Whitney stratifiable if its graph admits a C^∞-Whitney stratification.

The critical point notion for Lipschitz functions from $\mathbb{R}^n$ to $\mathbb{R}$ is based on Clarke's subdifferentials.

Definition 43 (Clarke regular/critical for $\mathbb{R}^n \longrightarrow \mathbb{R}$; see [13, 10]). A point $\bar{x}$ is called Clarke critical for a Lipschitz function $f : \mathbb{R}^n \longrightarrow \mathbb{R}$ if $0 \in \partial f(\bar{x})$, where $\partial f(\bar{x})$ is Clarke's subdifferential of f at $\bar{x}$. Otherwise, $\bar{x}$ is called Clarke regular. A point $y \in \mathbb{R}^k$ is called a Clarke regular (resp. critical) value for f if $f^{-1}(y)$ contains no Clarke critical points (resp. contains at least one Clarke critical point).

Remark 34. We refer to [66] for the similar treatment of continuous selections of smooth functions.

Theorem 46 (Sard's Theorem for $\mathbb{R}^n \longrightarrow \mathbb{R}$ [10]). *The set of Clarke's critical values of a C^∞-Whitney stratifiable function $f : \mathbb{R}^n \longrightarrow \mathbb{R}$ has Lebesgue measure zero.*

The proof of Theorem 46 is based on projection formulas for Clarke's subdifferentials and involves the Whitney property of the corresponding stratification (see [10]).

We additionally refer to the recent work [51] for other Clarke-like notions of critical points for functions from $\mathbb{R}^n$ to $\mathbb{R}^k$ with stratifiable graphs.

Min-type functions and MFC

Let $F_1 := (F_{1,i}, i = 1,\ldots,k)^T$, $F_2 := (F_{2,i}, i = 1,\ldots,k)^T \in C^\infty(\mathbb{R}^n, \mathbb{R}^k)$. Setting $g_i := \min\{F_{1,i}, F_{2,i}\}$, $i = 1,\ldots,k$, we define the min-type function

$$G : \begin{cases} \mathbb{R}^n \longrightarrow & \mathbb{R}^k, \\ x \mapsto & (g_1(x),\ldots,g_k(x)). \end{cases} \tag{6.1}$$

For min-type functions, we define the notion of topologically regular/critical points as follows.

Definition 44 (Topologically regular/critical for min-functions). A point $\bar{x}$ is called topologically regular for a min-type function $G : \mathbb{R}^n \longrightarrow \mathbb{R}^k$ as in (6.1) if any k vectors $(w_1,\ldots,w_k) \in \partial g_1(\bar{x}) \times \cdots \times \partial g_1(\bar{x})$ are linearly independent, where $\partial g_i(\bar{x})$ is Clarke's subdifferential of g_i at $\bar{x}$. Otherwise, $\bar{x}$ is called topologically critical. A point $y \in \mathbb{R}^k$ is called a topologically regular (resp. critical) value for G if $G^{-1}(y)$ contains no topologically critical points (resp. contains at least one topologically critical point).

Remark 35 (Topologically regular for min-functions and MFC). Note that $\bar{x}$ is a topologically regular point if and only if the MFC holds at $\bar{x}$ (see Definition 7). Indeed,

$$\partial g_i(\bar{x}) = \partial \min\{F_{1,i}, F_{2,i}\}(\bar{x}) = \text{conv}\{\nabla F_{j,i}(\bar{x}) \mid F_{j,i}(\bar{x}) = g_i(x)\}.$$

In the case $k = 1$ of a single min-function, the notion of Clarke regular points according to Definition 43 coincides with that of topologically regular points according to Definition 44.

Theorem 47 (Sard's Theorem for min-functions). *The set of topologically critical values of a min-type function* $G : \mathbb{R}^n \longrightarrow \mathbb{R}^k$ *as in (6.1) has Lebesgue measure zero.*

Proof. Let $\mathcal{J}$ be a collection of all $\emptyset \neq J^i \subset \{1,2\}$, $i = 1,\ldots,k$. For each element $J = (J^i, i = 1,\ldots,k)$ from this collection $\mathcal{J}$, we define a C^∞-function

$$F^J : \begin{cases} \mathbb{R}^n \longrightarrow \qquad\qquad \mathbb{R}^{|J|}, \\ x \quad \mapsto \quad (F_{j,i}, j \in J^i, i = 1,\ldots,k), \end{cases}$$

where $|J| := \sum_{i=1}^{k} |J^i|$. Note that $k \leq |J| \leq 2k$.

Furthermore, if $\bar{x}$ is a topologically critical point for G (see Definition 44) then $\bar{x}$ is critical for F^J (in the classical sense, see Definition 37) with

$$J := (J^i, i = 1,\ldots,k), J^i = \{ j \in \{1,2\} \,|\, F_{j,i}(x) = g_i(x) \}$$

and

$$F_{j,i}(x) = g_i(x) \text{ for all } j \in J^i. \tag{6.2}$$

Hence, applying the classical Sard's Theorem for F^J, $J \in \mathcal{J}$, we get the desired result. Note that the collection $\mathcal{J}$ is finite and that critical points of G produce the critical values of F^J with the same components indexed by $j \in J^i$ from (6.2). $\square$

6.3 Regularity and implicit functions

We discuss different notions of metric and topological regularity/criticality introduced in Section 6.2 for various nonsmooth functions. We show that these notions naturally generalize important consequences of the regularity property in the smooth setting,

$$DF(\bar{x})[\mathbb{R}^n] = \mathbb{R}^k \text{ for } F \in C^\infty(\mathbb{R}^n, \mathbb{R}^k).$$

Indeed, metric regularity corresponds to the well-known Lyusternik-Graves Theorem, and topological regularity (at least for min-type functions) resembles transversality and the implicit function theorem. We show that the concepts of metric and topological regularity do not already coincide for min-type functions. This gives rise to establishing nonsmooth analysis along the lines of topological regularity based mainly on the application of implicit function theorems (close to the very well developed nonsmooth analysis based on metric regularity; see [49, 81, 94, 104]). It is a very challenging issue to apply the ideas behind topological regularity for different kinds of nonsmooth functions and to get its analytical description. Note that the definition of topological regularity has been given only for min-functions up to now (see Definition 44) and is written in terms of Clarke's subdifferentials.

Consequences of regularity for smooth functions

We formulate Lyusternik-Graves Theorem as follows.

Theorem 48 (Lyusternik-Graves Theorem [30, 49, 89]). *Let $F \in C^1(\mathbb{R}^n, \mathbb{R}^k)$ and $\bar{x} \in \mathbb{R}^n$ be regular (according to Definition 37). Then, there exists $K > 0$ such that*

(a) $B(F(x), t) \subset F(B(x, Kt))$ for x close to $\bar{x}$ and small $t > 0$,
(b) $dist(x, F^{-1}(y)) \leq K dist(F(x), y)$ for (x, y) close to $(\bar{x}, F(\bar{x}))$.

Remark 36. It can be immediately seen that part (b) from Theorem 48 is exactly the definition of metric regularity (see Definition 40). Hence, in the smooth case metric regularity is a consequence of the property

$$DF(\bar{x})[\mathbb{R}^n] = \mathbb{R}^k.$$

It is worth mentioning that the standard statement of Lyusternik is

$$T_{\bar{x}}F^{-1}(F(\bar{x})) = \{\xi \in \mathbb{R}^n \mid DF(\bar{x}) \cdot \xi = 0\},$$

where $T_{\bar{x}}F^{-1}(F(\bar{x}))$ is the tangent space of the level set $F^{-1}(F(\bar{x}))$ at $\bar{x}$.

Remark 36 gives rise to linking regularity with results from transversality theory (see, e.g., [63]).

Definition 45 (Transversality of manifolds). Let M_1 and M_2 be manifolds in $\mathbb{R}^n$. We say that M_1 and M_2 intersect transversally if at every point $\bar{x} \in M_1 \cap M_2$ the following condition on the tangent spaces holds:

$$T_{\bar{x}}M_1 + T_{\bar{x}}M_2 = \mathbb{R}^n.$$

Definition 46 (Transversality of mappings). Let M be a manifold in $\mathbb{R}^k$ and $F \in C^1(\mathbb{R}^n, \mathbb{R}^k)$. We say that F meets M transversally if the following two manifolds M_1, M_2 in $\mathbb{R}^n \times \mathbb{R}^k$ intersect transversally:

$$M_1 := \operatorname{graph}(F), M_2 := \mathbb{R}^n \times M.$$

Note that F meets M transversally if and only if at every $x \in \mathbb{R}^n$ with $f(x) \in M$ the following holds:
$$DF(\bar{x})[\mathbb{R}^n] + T_{f(x)}M = \mathbb{R}^k.$$

Hence, $y \in \mathbb{R}^k$ is a regular value for F (according to Definition 37) if and only if F meets $\{y\}$ transversally.

Furthermore, if F meets M (a manifold of codimension m) transversally, then either $F^{-1}(M) = \emptyset$ or otherwise $F^{-1}(M)$ is a manifold in $\mathbb{R}^n$ of codimension m. Moreover, for $\bar{x} \in F^{-1}(M)$, we have

$$T_{\bar{x}}F^{-1}(M) = DF(\bar{x})^{-1}T_{F(\bar{x})M}.$$

From the considerations above, we get the following result.

Theorem 49 (Regular values and manifold). *Let $\bar{y}$ be a regular value of $F \in C^1(\mathbb{R}^n, \mathbb{R}^k)$. Then, $F^{-1}(\bar{y}) \neq \emptyset$ is a manifold in $\mathbb{R}^n$ of dimension $n - k$ and*

$$T_{\bar{x}}F^{-1}(\bar{y}) = \{\xi \in \mathbb{R}^n \mid DF(\bar{x}) \cdot \xi = 0\} \text{ for all } \bar{x} \in F^{-1}(\bar{y}).$$

We point out that the proof of Theorem 49 is mainly based on the application of the classical implicit function theorem. Indeed, we need to show that $F^{-1}(\bar{y})$ is locally diffeomorphic to $\mathbb{R}^{n-k}$ (i.e., is a manifold of dimension $n - k$). For that, we parameterize the solution set of $F(x) = \bar{y}$ using the implicit function theorem, which can be applied due to the fact that each $\bar{x} \in F^{-1}(\bar{y})$ is a regular point for F.

On metric and topological regularity for min-type functions

Theorems 48 and 49 give rise to concentrating either on metric or topological properties of $F^{-1}(\bar{y})$ for a possibly nonsmooth function $F : \mathbb{R}^n \longrightarrow \mathbb{R}^k$, $\bar{y} \in \mathbb{R}^k$. Indeed, taking min-type functions into consideration, we obtain the following result (similar to that of Theorem 49 in the smooth setting).

Theorem 50 (Topological regularity and Lipschitz manifold). *Let $\bar{y}$ be a topologically regular value (see Definition 44) of a min-type function $G : \mathbb{R}^n \longrightarrow \mathbb{R}^k$ given as in (6.1). Furthermore, assume the conjectured equivalence of the MFC and SMFC (see Definition 7). Then, $F^{-1}(\bar{y})$ is a Lipschitz manifold in $\mathbb{R}^n$ of dimension $n - k$.*

Proof. It follows immediately from Corollary 2, Definition 44, and Remark 35. $\square$

It is worth mentioning that metrically and topologically regular values do not already coincide in the setting of min-type functions.

Remark 37 (Metrically and topologically regular values). From Propositions 3 and 4, each topologically regular value is also metrically regular, but not vice versa. In fact, let

$$G : \begin{cases} \mathbb{R}^3 \longrightarrow & \mathbb{R}^2, \\ x \mapsto & (\min\{x,y\}, \min\{x+y-\sqrt{2}z, x+y+\sqrt{2}z\}). \end{cases}$$

From Example 6, we see that 0 is a metrically regular value for G but not topologically regular.

Finally, we point out that topological considerations of $F^{-1}(\bar{y})$ for general nonsmooth functions $F : \mathbb{R}^n \longrightarrow \mathbb{R}^k$ should involve conditions that guarantee that $F^{-1}(\bar{y})$ is a Lipschitz manifold of the right dimension. It requires an application of nonsmooth versions of implicit function theorems (e.g., due to Clarke [13] or Kummer [85, 86]). Certainly, analytical descriptions of the latter application depends heavily on the nonsmoothness type of F. For min-type functions, we refer to Definition 44. Min-max functions might be handled using results on their Clarke subdifferentials from Section 3.2.2. In general, topological properties of $\bar{y}$-level sets of a nonsmooth function $F : \mathbb{R}^n \longrightarrow \mathbb{R}^k$ are a very challenging issue.

Appendix A
Topology

We briefly recall some notions from general topology. Those are needed for deformation and cell attachment of topological spaces as used in the critical point theory. The strong or Whitney topology is discussed. Note that the stability and genericity results for constraint qualifications, reduction ansatz and nondegeneracy of stationary points from Chapters 2–5, are stated w.r.t. this Whitney topology.

A.1 Cell attachment and deformation

For stating cell-attachment and deformation results within the scope of critical point theory (see Theorems 1 and 2) we need some notions from general topology (e.g., [63, 113]).

Definition 47 (Homotopy of mappings). Let X and Y be topological spaces.

(a) The continuous mappings $h_i : X \longrightarrow Y$, $i = 0, 1$, are called homotopic (notation $h_0 \simeq h_1$) if a continuous $H : [0, 1] \times X \longrightarrow Y$ exists such that

$$H(0, \cdot) = h_0(\cdot) \text{ and } H(1, \cdot) = h_1(\cdot).$$

  H is called a homotopy between h_0 and h_1.

(b) Let $h_0 \simeq h_1$ and $h_0 |_A = h_1 |_A$ for $A \subset X$. If, moreover,

$$H(t, x) = h_0(x) \text{ for all } (t, x) \in [0, 1] \times A,$$

  then h_0 and h_1 are called homotopic relative to A (notation $h_0 \simeq h_1$ rel A).

Definition 48 (Homotopy of topological spaces). The topological spaces X and Y are called homotopy-equivalent (notation $X \simeq Y$) if continuous mappings $g : X \longrightarrow Y$ and $h : Y \longrightarrow X$ exist such that

$$h \circ g \simeq Id_X \text{ and } g \circ h \simeq Id_Y.$$

We say also that X and Y have the same homotopy type if $X \simeq Y$.

The following definition clarifies the meaning of what we call "deformation."

Definition 49 (Deformation retract). Let X be a topological space and $A \subset X$ be endowed with the relative topology induced from X. We denote by $i : A \longrightarrow X$ the inclusion mapping (i.e., $i(a) = a$ for all $a \in A$).

(i) A is called a retract of X if a continuous mapping $r : X \longrightarrow A$ exists such that $r \circ i = Id_A$; we call r a retraction.

(ii) A is called a deformation retract of X if a retraction $r : X \longrightarrow A$ exists such that $i \circ r \simeq Id_X$.

(iii) A is called a strong deformation retract of X if a retraction $r : X \longrightarrow A$ exists such that $i \circ r \simeq Id_X$ rel A.

Now we turn our attention to the so-called cell-attachment procedure.

Definition 50 (Quotient space). Let X be a topological space, and $R \subset X \times X$ be an equivalence relation on X. The equivalence class of $x \in X$ under R is denoted by $\langle x \rangle$ and the set of all equivalence classes by X/R. Let $p : X \longrightarrow X/R$ be the projection mapping that sends each element of X to its equivalence class. The quotient topology on X/R is the finest topology for which q is continuous. Equivalently, a set of equivalence classes in X/R is open if and only if their union is open in X. The set X/R endowed with the quotient topology is called a quotient space.

Definition 51 (Attachment of topological spaces). Let X and Y be topological spaces, A a closed subset of X, and $f : A \longrightarrow Y$ a continuous mapping. Let R be the equivalence relation on the topological sum $X \cup Y$ induced by $a \sim_R f(a)$ for all $a \in A$. The quotient space $Y \cup_f X := (X \cup Y)/R$ is said to be an attachment of X on Y by means of f. One says that the points a, a' with $f(a) = f(a')$ are identified under the relation R.

Definition 52 (Cell attachment). Let X be a topological space. $D^k := \{x \in \mathbb{R}^k \mid \|x\| \leq 1\}$ denotes the closed unit ball in $\mathbb{R}^k$, $k \geq 1$, with its boundary $S^{k-1} := \{x \in \mathbb{R}^k \mid \|x\| = 1\}$. Let $\phi : S^{k-1} \longrightarrow X$ be a continuous mapping. The topological space $X \cup_\phi D^k$ is said to be a k-cell attachment on X by means of ϕ.

A.2 Whitney topology

We introduce a suitable topology on the space $C^k(\mathbb{R}^n, \mathbb{R})$ that takes the asymptotic behavior of functions into account. We follow mainly [63].

Let $\alpha = (\alpha_1, \ldots, \alpha_n) \in \mathbb{N}^n$, $|\alpha| = \sum_{i=1}^{n} \alpha_i$, and denote by ∂^α the α-th partial derivative of $f \in C^k(\mathbb{R}^n, \mathbb{R})$ and $|\alpha| \leq k$. Put

$$C_+(\mathbb{R}^n, \mathbb{R}) := \{\phi : \mathbb{R}^n \longrightarrow \mathbb{R} \mid \phi \text{ continuous and } \phi(x) > 0 \text{ for all } x \in \mathbb{R}^n\}.$$

For fixed $k \in \mathbb{N}$ and $(\phi, f) \in C_+(\mathbb{R}^n, \mathbb{R}) \times C^k(\mathbb{R}^n, \mathbb{R})$, we define the set

$$V^k_{\phi,f} := \left\{ g \in C^k(\mathbb{R}^n, \mathbb{R}) \mid |\partial^\alpha f(x) - \partial^\alpha g(x)| < \phi(x) \text{ for all } x \in \mathbb{R}^n, \atop \text{for all } \alpha \text{ with } |\alpha| \leq k \right\}.$$

Definition 53 (Whitney topology; see, e.g., [63, 79]). For fixed $k \in \mathbb{N}$, the set system

$$\mathscr{V} := \left\{ V^k_{\phi,f}, \ (\phi, f) \in C_+(\mathbb{R}^n, \mathbb{R}) \times C^k(\mathbb{R}^n, \mathbb{R}) \right\}$$

forms a basis for the unique topology on $C^k(\mathbb{R}^n, \mathbb{R})$, meaning

(i) $$\bigcup_{V^k_{\phi,f} \in \mathscr{V}} V^k_{\phi,f} = C^k(\mathbb{R}^n, \mathbb{R})$$

and

(ii) for every $h \in V^k_{\phi,f} \cap V^k_{\psi,g}$ there exists $\xi \in C_+(\mathbb{R}^n, \mathbb{R})$ such that $V^k_{\xi,h} \subset V^k_{\phi,f} \cap V^k_{\psi,g}$.

This topology is the intersection of all topologies on $C^k(\mathbb{R}^n, \mathbb{R})$ containing $\mathscr{V}$. It is called the strong or Whitney C^k-topology and is denoted by C^k_s.

In a straightforward way, the C^l_s-topology for $C^k(\mathbb{R}^n, \mathbb{R})$, where $l \leq k$, is defined, as well as the C^k_s-topology for $C^\infty(\mathbb{R}^n, \mathbb{R})$. The C^k_s-topology for a finite product of function spaces is defined to be the product topology.

We mention some useful properties of the Whitney topology.

Remark 38 (Asymptotic behavior and Whitney topology [63]). The reason for introducing the C^k_s-topology comes from the fact that this topology takes the asymptotic behavior of functions into account. For example, a function $f \in C^2(\mathbb{R}^n, \mathbb{R})$ is called nondegenerate if all its critical points are nondegenerate. Obviously, f is nondegenerate if and only if $\|Df(x)\| + |\det D^2 f(x)| > 0$ for all $x \in \mathbb{R}^n$. From this observation, it follows that the subset of $C^2(\mathbb{R}^n, \mathbb{R})$ consisting of all nondegenerate functions is C^2_s-open.

Remark 39 (Baire space and Whitney topology [63]). Note that $C^k(\mathbb{R}^n, \mathbb{R})$ endowed with the C^k_s-topology is a Baire space, meaning every countable intersection of C^k_s-open and -dense subsets of $C^k(\mathbb{R}^n, \mathbb{R})$ is C^k_s-dense in $C^k(\mathbb{R}^n, \mathbb{R})$. We say that a set is C^k_s-generic if it contains a countable intersection of C^k_s-open and C^k_s-dense subsets. Hence, generic sets in a Baire space are dense as well.

Appendix B
Analysis

We recall the notions of smooth, Lipschitz, and topological manifolds. The nonsmooth versions of inverse and implicit function theorems due to Clarke and Kummer are stated. Finally, we present some elements of the transversality theory, in particular, transversality of manifolds and mappings. The Thom's transversality theorem and its jet version are stated. Note that the genericity results from Chapters 2–5 are obtained mainly by the application of the Thom's transversality theorem.

B.1 Manifolds and implicit functions

We recall the definitions of topological, Lipschitz, and C^k-manifolds.

Definition 54 (Manifolds [63, 103]). A subset $\mathscr{M} \subseteq \mathbb{R}^n$ is called a topological (resp. Lipschitz or C^k-) manifold (with boundary) of dimension $m \geq 0$ if for each $\bar{x} \in \mathscr{M}$ there exist open neighborhoods $U \subseteq \mathbb{R}^n$ of $\bar{x}$ and $V \subseteq \mathbb{R}^n$ of 0 and a homeomorphism $H : U \to V$ (resp. with H, H^{-1} being Lipschitz continuous or H being a C^k-diffeomorphism) such that

(i) $H(\bar{x}) = 0$

and

(ii)   either

$$H(\mathscr{M} \cap U) = (\mathbb{R}^m \times \{0_{n-m}\}) \cap V$$

or

$$H(\mathscr{M} \cap U) = (\mathbb{H} \times \mathbb{R}^{m-1} \times \{0_{n-m}\}) \cap V.$$

In the latter case, $\bar{x}$ is said to be a boundary point of $\mathscr{M}$.

If for all $x \in \mathscr{M}$ the first case in (ii) holds, then $\mathscr{M}$ is called a topological (resp. Lipschitz or C^k-) manifold of dimension m.

The parameterization of manifolds may usually be constructed via implicit functions. We recapitulate the classical implicit function theorem and its nonsmooth versions due to Clarke and Kummer.

Theorem 51 (Classical inverse function theorem; e.g., [59]). *Let $F : \mathbb{R}^n \longrightarrow \mathbb{R}^n$ be k-times continuously differentiable near $\bar{x}$. If $DF(\bar{x})$ is nonsingular, then F has the unique k-times continuously differentiable inverse function F^{-1} locally around $\bar{x}$.*

Theorem 52 (Classical implicit function theorem; e.g., [59]). *Let $G : \mathbb{R}^{n-k} \times \mathbb{R}^k \longrightarrow \mathbb{R}^k$ be k-times continuously differentiable near $(\bar{y}, \bar{z}) \in \mathbb{R}^{n-k} \times \mathbb{R}^k$ with $G(\bar{y}, \bar{z}) = 0$. Suppose that $D_z G(\bar{y}, \bar{z})$ is a nonsingular matrix. Then there exist an $\mathbb{R}^{n-k}$-neighborhood Y of $\bar{y}$, an $\mathbb{R}^k$-neighborhood Z of $\bar{z}$, and a k-times continuously differentiable function $\zeta : Y \longrightarrow Z$ such that $\zeta(\bar{y}) = \bar{z}$ and for every $(y, z) \in Y \times Z$ it holds that*

$$G(y, z) = 0 \text{ if and only if } z = \zeta(y).$$

For a vector-valued function $G = (g_1, \dots, g_k) : \mathbb{R}^n \longrightarrow \mathbb{R}^k$ with g_i being Lipschitz near $\bar{x} \in \mathbb{R}^n$, the set

$$\partial G(\bar{x}) := \text{conv}\{\lim DG(x_i) \,|\, x_i \longrightarrow \bar{x}, x_i \notin \Omega_G\}$$

is called Clarke's generalized Jacobian, where $\Omega_G \subset \mathbb{R}^n$ denotes the set of points at which G fails to be differentiable.

Theorem 53 (Clarke's inverse function theorem [13]). *Let $F : \mathbb{R}^n \longrightarrow \mathbb{R}^n$ be Lipschitz near $\bar{x}$. If all matrices in $\partial F(\bar{x})$ are nonsingular, then F has the unique Lipschitz inverse function F^{-1} locally around $\bar{x}$.*

Theorem 54 (Clarke's implicit function theorem [13]). *Let $G : \mathbb{R}^{n-k} \times \mathbb{R}^k \longrightarrow \mathbb{R}^k$ be Lipschitz near $(\bar{y}, \bar{z}) \in \mathbb{R}^{n-k} \times \mathbb{R}^k$ with $G(\bar{y}, \bar{z}) = 0$. Suppose that*

$$\pi_z \partial G(\bar{y}, \bar{z}) := \{M \in \mathbb{R}^{k \times k} \,|\, \text{there exists } N \in \mathbb{R}^{k \times n} \text{ with } [N, M] \in \partial G(\bar{y}, \bar{z})\}$$

is of maximal rank (i.e., it contains merely nonsingular matrices). Then there exist an $\mathbb{R}^{n-k}$-neighborhood Y of $\bar{y}$, an $\mathbb{R}^k$-neighborhood Z of $\bar{z}$, and a Lipschitz function $\zeta : Y \longrightarrow Z$ such that $\zeta(\bar{y}) = \bar{z}$ and for every $(y, z) \in Y \times Z$ it holds that

$$G(y, z) = 0 \text{ if and only if } z = \zeta(y).$$

For a vector-valued function $G = (g_1, \dots, g_k) : \mathbb{R}^n \longrightarrow \mathbb{R}^k$, the mapping $TG(\bar{x}) : \mathbb{R}^n \longrightarrow \mathbb{R}^k$ with

$$TG(\bar{x})(\bar{u}) := \left\{ v \in \mathbb{R}^k \,\middle|\, \begin{array}{l} v = \lim_{k \to \infty} \dfrac{f(x_k + t_k u_k) - f(x_k)}{t_k} \\ \text{for certain } t_k \downarrow 0, \, x_k \longrightarrow \bar{x}, u_k \longrightarrow \bar{u} \end{array} \right\}$$

is called a Thibault derivative at $\bar{x}$ (see [114, 115]) or a strict graphical derivative (see [104]).

If, additionally, g_i are Lipschitz near $\bar{x} \in \mathbb{R}^n$, then we may omit the sequence $u_k \longrightarrow \bar{u}$ in the definition of $TG(\bar{x})(\bar{u})$, and we get

$$TG(\bar{x})(\bar{u}) = \left\{ v \in \mathbb{R}^k \,\middle|\, \begin{array}{l} v = \lim\limits_{k \to \infty} \dfrac{f(x_k + t_k \bar{u}) - f(x_k)}{t_k} \\ \text{for certain } t_k \downarrow 0,\, x_k \longrightarrow \bar{x} \end{array} \right\}.$$

Necessary and sufficient conditions for local invertability of Lipschitz functions can be given in terms of Thibault derivatives.

Theorem 55 (Kummer's inverse function theorem [81, 86]). *Let* $F : \mathbb{R}^n \longrightarrow \mathbb{R}^n$ *be Lipschitz near* $\bar{x}$. *Then the following statements are equivalent:*

(i)   *F has the locally unique Lipschitz inverse function F^{-1}.*

(ii)   *There exists $c > 0$ such that*

$$\|F(x) - F(x')\| \geq c\|x - x'\| \text{ for all } x,\, x' \text{ with } \|\bar{x} - x\| \leq c, \|\bar{x} - x'\| \leq c.$$

(iii)   *$TF(\bar{x})$ is injective (i.e., $0 \notin TF(\bar{x})(u)$ for all $u \neq 0$).*

Theorem 56 (Kummer's implicit function theorem [81, 85]). *Let* $G : \mathbb{R}^{n-k} \times \mathbb{R}^k \longrightarrow \mathbb{R}^k$ *be Lipschitz near* $(\bar{y}, \bar{z}) \in \mathbb{R}^{n-k} \times \mathbb{R}^k$ *with* $G(\bar{y}, \bar{z}) = 0$. *Then the following statements are equivalent:*

(i)   *There exist $\mathbb{R}^{n-k}$-neighborhoods Y of $\bar{y}$ and W of 0, an $\mathbb{R}^k$-neighborhood Z of $\bar{z}$, and a Lipschitz function $\zeta : Y \times W \longrightarrow Z$ such that $\zeta(\bar{y}, 0) = \bar{z}$ and for every $(y, z, w) \in Y \times Z \times W$ it holds that*

$$G(y, z) = w \text{ if and only if } z = \zeta(y, w).$$

(ii)   *$0 \notin TG(\bar{y}, \bar{z})(0, u)$ for all $u \neq 0$.*

Note that the injectivity of $TF(\bar{x})$ in Theorem 55 is in general weaker than Clarke's requirement that all matrices in $\partial F(\bar{x})$ be nonsingular. In fact, there exists a Lipschitz homeomorphism F of $\mathbb{R}^2$ such that $\partial F(\bar{x})$ contains the zero matrix (see Example BE.3 in [81]). We point out that Theorem 56 gives a necessary and sufficient condition for the existence of implicit functions. Recall that Clarke's IFT (see Theorem 54) gives only a sufficient condition for that fact. For other versions of implicit function theorems, see [18].

B.2 Transversality

We briefly review the main notions and results of transversality theory, in particular Thom's transversality theorem.

Definition 55 (Transversality of manifolds; e.g., [63]). Let M_1 and M_2 be manifolds in $\mathbb{R}^n$. We say that M_1 and M_2 intersect transversally (notation $M_1 \pitchfork M_2$) if at every point $\bar{x} \in M_1 \cap M_2$ the following condition on the tangent spaces holds:

$$T_{\bar{v}}M_1 + T_{\bar{v}}M_2 = \mathbb{R}^n.$$

Definition 56 (Transversality of mappings; e.g., [63]). Let M be a manifold in $\mathbb{R}^k$ and $F \in C^1(\mathbb{R}^n, \mathbb{R}^k)$. We say that F meets M transversally (notation $F \pitchfork M$) if the following two manifolds M_1, M_2 in $\mathbb{R}^n \times \mathbb{R}^k$ intersect transversally:

$$M_1 := \text{graph}(F), \quad M_2 := \mathbb{R}^n \times M.$$

Note that F meets M transversally if and only if at every $x \in \mathbb{R}^n$ with $f(x) \in M$ the following holds:

$$DF(\bar{x})[\mathbb{R}^n] + T_{f(x)}M = \mathbb{R}^k.$$

Furthermore, if F meets M (a manifold of codimension m) transversally, then either $F^{-1}(M) = \emptyset$ or otherwise $F^{-1}(M)$ is a manifold in $\mathbb{R}^n$ of codimension m (see [63]). The latter fact is often used for the conclusion as follows. If $F \in C^1(\mathbb{R}^n, \mathbb{R}^k)$ meets a manifold $M \subset \mathbb{R}^k$ transversally and the codimension of M is greater than n, then $F^{-1}(M) = \emptyset$. We refer to this as a "counting dimension and codimension" argument.

The following result is a key point of transversality theory.

Theorem 57 (Thom's transversality theorem; e.g., [63]). *Let M be a manifold in $\mathbb{R}^k$, and denote*

$$\pitchfork M := \Big\{ F \in C^\infty(R^n, \mathbb{R}^k) \,|\, F \pitchfork M \Big\}.$$

Then, $\pitchfork M$ is C_s^m-dense for all m; moreover, if M is closed as a subset of $\mathbb{R}^k$, then $\pitchfork M$ is C_s^m-open for all $m \geq 1$.

The proof of Thom's transversality theorem is mainly based on the application of the classical Sard's Theorem.

Definition 57 (Regular/critical points; e.g., [63]). Let $F \in C^\infty(\mathbb{R}^n, \mathbb{R}^k)$. A point $\bar{x} \in \mathbb{R}^n$ is called critical if the linear map from $\mathbb{R}^n$ to $\mathbb{R}^k$ given by $\xi \mapsto DF(\bar{x})\xi$ is not surjective. In other words, $\bar{x}$ is regular (i.e., not critical) if and only if $DF(\bar{x})[\mathbb{R}^n] = \mathbb{R}^k$. A point $y \in \mathbb{R}^k$ is called a regular (resp. critical) value for F if $F^{-1}(y)$ contains no critical points (resp. contains at least one critical point).

Theorem 58 (Sard's Theorem; e.g., [63]). *The set of critical values of $F \in C^\infty(\mathbb{R}^n, \mathbb{R}^k)$ has Lebesgue measure zero.*

Finally, we present a generalization of Thom's transversality theorem, namely, the (structured) jet transversality theorem.

Definition 58 (Jet-extension). Let $F \in C^\infty(\mathbb{R}^n, \mathbb{R}^k)$. The l-jet extension $j^l F$ of F is defined to be a mapping,

$$j^l F : x \longrightarrow \left[x, F(x), \frac{\partial}{\partial x_1}F_1, \ldots, \frac{\partial}{\partial x_n}F_k, \ldots, \frac{\partial^l}{\partial x_n \partial x_n \ldots \partial x_n}F_k \right],$$

where the partial derivatives are listed according to the "order-convention." Let $N(n,k,l)$ denote the length of the vector $j^l f(x)$. The jet space $\mathscr{J}(n,k,l)$ is defined to be the space $\mathbb{R}^{N(n,k,l)}$.

Theorem 59 (Jet transversality theorem; e.g., [63]). *Let $\mathcal{M} \subset \mathcal{J}(n,k,l)$ be a manifold, and denote*

$$\pitchfork^l \mathcal{M} := \left\{ F \in C^\infty(R^n, \mathbb{R}^k) \mid j^l F \pitchfork \mathcal{M} \right\}.$$

Then, $\pitchfork \mathcal{M}$ is C_s^m-dense for all m; moreover, if $\mathcal{M}$ is closed as a subset of $\mathcal{J}(n,k,l)$, then $\pitchfork \mathcal{M}$ is C_s^m-open for all $m \geq 1$.

We also mention the structured jet transversality theorem from [33]. In addition to Theorem 59, Whitney stratified manifolds come into play here. Moreover, reduced jet extensions model the possibility of different function dependences on any variable involved. Furthermore, the multijet transversality theorem should be mentioned (see [63]). There, the possibility of certain regular behavior of functions w.r.t. any subset of points is considered.

References

1. W. ACHTZIGER, C. KANZOW, *Mathematical programs with vanishing constraints: optimality conditions and constraint qualifications,* Mathematical Programming, Vol. 114, No. 1, pp. 69–99, 2008.
2. A. A. AGRACHEV, D. PALLASCHKE, S. SCHOLTES, *On Morse theory for piecewise smooth functions,* Journal of Dynamical and Control Systems, Vol. 3, No. 4, pp. 449–469, 1997.
3. V.I. ARNOLD, *Singularity Theory,* Cambridge University Press, Cambridge, 1981.
4. A.V. ARUTYUNOV, *Optimality Conditions: Abnormal and Degenerate Problems,* Kluwer Academic Publishers, Dordrecht, 2000.
5. J.F. BARD, *Practical Bilevel Optimization: Algorithms and Applications,* Kluwer Academic Publishers, Dordrecht, 1998.
6. S.G. BARTELS, L. KUNTZ, S. SCHOLTES, *Continuous selections of linear functions and nonsmooth critical point theory,* Nonlinear Analysis: Theory, Methods & Applications, Vol. 24, No. 3, pp. 287–307, 1995.
7. A. BEN-TAL, L. EL GHAOUI, A. NEMIROVSKI, *Robust Optimization,* Princeton University Press, Princeton, New Jersey, 2009.
8. J. BOCHNAK, M. COSTE, M.-F. ROY, *Real Algebraic Geometry,* Springer, New York, 1998.
9. J.F. BONNANS, A. SHAPIRO, *Perturbation Analysis of Optimization Problems,* Springer, New York, 2000.
10. J. BOLTE, A. DANIILIDIS, A. LEWIS, M. SHIOTA, *Clarke subgradients of stratifiable functions,* SIAM Journal on Optimization, Vol. 18, No. 2, pp. 556–572, 2007.
11. P. BRÖCKER, L. LANDER, *Differentiable Germs and Catastrophes,* Cambridge University Press, Cambridge, 1975.
12. ST. BÜTIKOFER, D. KLATTE, *A nonsmooth Newton method with path search and its use in solving $C^{1,1}$ programs and semi-infinite problems,* SIAM Journal on Optimization, Vol. 20, No. 5, pp. 2381–2412, 2010.
13. F.H. CLARKE, *Optimization and Nonsmooth Analysis,* Canadian Mathematical Society Series of Monographs and Advanced Texts, Wiley New York, 1983.
14. M. COSTE, *An Introduction to o-minimal Geometry,* Institut de Recherche Mathématique de Rennes, 1999.
15. S. DEMPE, *Foundations of Bilevel Programming,* Kluwer Academic Publishers, Dordrecht, 2002.
16. S. DEMPE, J. DUTTA, *Is bilevel programming a special case of a mathematical program with complementarity constraints,* Mathematical Programming, to appear.
17. S. DEMPE, H. GÜNZEL, H.TH. JONGEN, *On reducibility in bilevel problems,* SIAM Journal on Optimization, Vol. 20, No. 2, pp. 718–727, 2009.
18. A.L. DONTCHEV, R.T. ROCKAFELLAR, *Implicit Functions and Solution Mappings,* Springer, Dordrecht, 2009.

19. D. DORSCH, H. TH. JONGEN, V. SHIKHMAN, *On topological properties of min-max functions,* Set-Valued and Variational Analysis, Vol. 19, No. 2, pp. 237–253, 2011.

20. D. DORSCH, V. SHIKHMAN, O. STEIN, *MPVC: critical point theory,* Journal of Global Optimization, to appear.

21. L. VAN DEN DRIES, *Tame Topology and O-minimal Structures,* Cambridge University Press, Cambridge, 1998.

22. F. FACCHINEI, H. JIANG, L. QI, *A smoothing method for mathematical programs with equilibrium constraints,* Mathematical Programming, Series A, Vol. 85, No. 1, pp. 107–134, 1999.

23. F. FACCHINEI, J.-S. PANG, *Finite-Dimensional Variational Inequalities and Complementarity Problems I, II,* Springer, Berlin, 2003.

24. M.L. FLEGEL, C. KANZOW, *Optimality conditions for mathematical programs with equilibrium constraints: Fritz John and Abadie-type approaches,* Journal of Optimization Theory and Applications, Vol. 124, No. 3, pp. 595–614, 2005.

25. M.L. FLEGEL, C. KANZOW, J.V. OUTRATA, *Optimality conditions for disjunctive programs with application to mathematical programs with equilibrium constraints,* Set-Valued Analysis, Vol. 15, No. 2, pp. 139–162, 2007.

26. C.A. FLOUDAS, H.TH. JONGEN, *Global optimization: local minima and transition points,* Journal of Global Optimization, Vol. 32, No. 3, pp. 409–415, 2005.

27. C.G. GIBSON, *Singular Points of Smooth Mappings,* Research Notes in Mathematics, Vol. 25, Pitman, London, 1979.

28. M. GOLUBITSKY, V. GUILLEMIN, *Stable Mappings and Their Singularities,* Graduate Texts in Mathematics, Springer, New York, 1979.

29. M. GORESKY, R. MACPHERSON, *Stratified Morse Theory,* Springer, New York, 1988.

30. L.M. GRAVES, *Some mapping theorems,* Duke Mathematical Journal, Vol. 17, pp. 114–117, 1950.

31. J. GUDDAT, H. TH. JONGEN, *Structural stability and nonlinear optimization,* Optimization, Vol. 18, No. 5, pp. 617–631, 1987.

32. J. GUDDAT, H.TH. JONGEN, J.-J. RÜCKMANN, *On stability and stationary points in nonlinear optimization,* Journal of the Australian Mathematical Society, Series B, Vol. 28, pp. 36–56, 1986.

33. H. GÜNZEL, *The structured jet transversality theorem,* Optimization, Vol. 57, No. 1, pp. 159–164, 2008.

34. H. GÜNZEL, F. GUERRA-VÁZQUEZ, H. TH. JONGEN, *Critical value functions have finite modulus of concavity,* SIAM Journal on Optimization, Vol. 16, No. 4, pp. 1044–1053, 2006.

35. H. GÜNZEL, H.TH. JONGEN, *Strong stability implies Mangasarian-Fromovitz constraint qualification,* Optimization, Vol. 55, pp. 605–610, 2006.

36. H. GÜNZEL, H.TH. JONGEN, O. STEIN, *On the closure of the feasible set in generalized semi-infinite programming,* Central European Journal of Operations Research, Vol. 15, No. 3, pp. 271–280, 2007.

37. H. GÜNZEL, H.TH. JONGEN, O. STEIN, *Generalized semi-infinite programming: the symmetric reduction ansatz,* Optimization Letters, Vol. 2, No. 3, pp. 415–424, 2008.

38. H. GÜNZEL, H.TH. JONGEN, O. STEIN, *Generalized semi-infinite programming: on generic local minimizers,* Journal of Global Optimization, Vol. 42, No. 3, pp. 413–421, 2008.

39. F. GUERRA-VÁZQUEZ, H. TH. JONGEN, V. SHIKHMAN, *General semi-infinite programming: symmetric Mangasarian Fromovitz constraint qualification and the closure of the feasible set,* SIAM Journal on Optimization, Vol. 20, No. 5, pp. 2487–2503, 2010.

40. F. GUERRA-VÁZQUEZ, J.-J. RÜCKMANN, O. STEIN, G. STILL, *Generalized semi-infinite programming: a tutorial,* Journal of Computational and Applied Mathematics, Vol. 217, No. 2, pp. 394–419, 2008.

41. R. HETTICH, G. STILL, *Second order optimality conditions for generalized semi-infinite programming problems,* Optimization, Vol. 34, No. 3, pp. 195–211, 1995.

42. M.W. HIRSCH, *Differential Topology,* Springer, New York, 1976.

43. T. HOHEISEL, C. KANZOW, *First- and second-order optimality conditions for mathematical programs with vanishing constraints,* Applications of Mathematics, Vol. 52, pp. 495–514, 2007.

44. T. HOHEISEL, C. KANZOW, *Stationary conditions for mathematical programs with vanishing constraints using weak constraint qualifications,* Journal of Mathematical Analysis and Applications, Vol. 337, No. 1, pp. 292–310, 2008.

45. T. HOHEISEL, C. KANZOW, *On the Abadie and Guignard constraint qualification for mathematical programs with vanishing constraints,* Optimization, Vol. 58, pp. 431–448, 2009.

46. T. HOHEISEL, C. KANZOW, J. OUTRATA, *Exact penalty results for mathematical programs with vanishing constraints,* Nonlinear Analysis: Theory, Methods and Applications, Vol. 72, No. 5, pp. 2514–2526, 2010.

47. T. HOHEISEL, C. KANZOW, A. SCHWARTZ, *Convergence of a local regularization approach for mathematical programs with complementarity or vanishing constraints,* Optimization Methods and Software, to appear.

48. A.D. IOFFE, *Nonsmooth analysis: differential calculus of nondifferentiable mappings,* Transactions of the American Mathematical Society, Vol. 266, pp. 1–56, 1981.

49. A.D. IOFFE, *Metric regularity and subdifferential calculus,* Russian Mathematical Surveys, Vol. 55, No. 3, pp. 501–558, 2000.

50. A.D. IOFFE, *A Sard theorem for tame set-valued mappings,* Journal of Mathematical Analysis and Applications, Vol. 335, No. 2, pp. 882–901, 2007.

51. A.D. IOFFE, *Critical values of set-valued maps with stratifiable graphs. Extensions of Sard and Smale-Sard theorems,* Proceedings of the American Mathematical Society, Vol. 136, pp. 3111–3119, 2008.

52. A.D. IOFFE, J.V. OUTRATA, *On metric and calmness qualification conditions in subdifferential calculus,* Set-Valued Analysis, Vol. 16, pp. 199–227, 2008.

53. A.F. IZMAILOV, *Mathematical programs with complementarity constraints: regularity, optimality conditions, and sensitivity,* Computational Mathematics and Mathematical Physics, Vol. 44, No. 7, pp. 1145–1164, 2004.

54. A.F. IZMAILOV, *Solution sensitivity for Karush-Kuhn-Tucker systems with non-unique Lagrange multipliers,* Optimization, Vol. 59, No. 5, pp. 747–775, 2010.

55. A.F. IZMAILOV, A.L. POGOSYAN, *Optimality conditions and Newton-type methods for mathematical programs with vanishing constraints,* Computational Mathematics and Mathematical Physics, Vol. 49, No. 7, pp. 1128–1140, 2009.

56. A.F. IZMAILOV, M.V. SOLODOV, *Mathematical programs with vanishing constraints: optimality conditions, sensitivity, and a relaxation method,* Journal of Optimization Theory and Applications, Vol. 142, No. 3, pp. 501–532, 2009.

57. A.F. IZMAILOV, M.V. SOLODOV, *An active set Newton method for mathematical programs with complementarity constraints,* SIAM Journal on Optimization, Vol. 19, No. 3, pp. 1003–1027, 2008.

58. A.F. IZMAILOV, A.A. TRETIAKOV, *2-Regular Solutions of Nonlinear Problems,* Literature in Physics and Mathematics, Moscow, 1999 (in Russian).

59. H.TH. JONGEN, P.G. SCHMIDT, *Analysis,* Wissenschaftsverlag Mainz, Aachen, 1988.

60. H.TH. JONGEN, *Parametric optimization: critical points and local minima,* Lectures in Applied Mathematics, Vol. 26, pp. 317–335, 1990.

61. H.TH. JONGEN, P. JONKER, F. TWILT, *Nonlinear Optimization in $\mathbb{R}^n$ I. Morse Theory, Chebyshev Approximation,* Peter Lang Verlag, Frankfurt am Main, 1983.

62. H.TH. JONGEN, P. JONKER, F. TWILT, *Nonlinear Optimization in $\mathbb{R}^n$ II. Transversality, Flows, Parametric Aspects,* Peter Lang Verlag, Frankfurt am Main., Bern, New York, 1986.

63. H.TH. JONGEN, P. JONKER, F. TWILT, *Nonlinear Optimization in Finite Dimensions,* Kluwer Academic Publishers, Dordrecht, 2000.

64. H. TH. JONGEN, P. JONKER, F. TWILT, *Critical sets in parametric optimization,* Mathematical Programming, Vol. 34, pp. 333–353, 1986.

65. H.TH. JONGEN, K. MEER, E. TRIESCH, *Optimization Theory,* Kluwer Academic Publishers, Dordrecht, 2004.

66. H.TH. JONGEN, D. PALLASCHKE, *On linearization and continuous selections of functions,* Optimization, Vol. 19, pp. 343–353, 1988.

67. H.TH. JONGEN, J.-J. RÜCKMANN, *On stability and deformation in semi-infinite optimization,* in: *Semi-Infinite Programming,* R. Reemtsen and Jan.-J. Rückmann (eds.), Kluwer Academic Publishers, Dordrecht, pp. 29–67, 1998.

68. H.TH. JONGEN, J.-J. RÜCKMANN, *One-parameter families of feasible sets in semi-infinite optimization,* Journal of Global Optimization, Vol. 14, No. 2, pp. 181–203, 1999.

69. H.TH. JONGEN, J.-J. RÜCKMANN, V. SHIKHMAN, *MPCC: critical point theory,* SIAM Journal on Optimization, Vol. 20, No. 1, pp. 473–484, 2009.

70. H.TH. JONGEN, JAN.-J. RÜCKMANN, V. SHIKHMAN, *On stability of the MPCC feasible set,* SIAM Journal on Optimization, Vol. 20, No. 3, pp. 1171–1184, 2009.

71. H.TH. JONGEN, J.-J. RÜCKMANN, O. STEIN, *Disjunctive optimization: critical point theory,* Journal of Optimization Theory and Applications, Vol. 93, No. 2, pp. 321–336, 1997.

72. H.TH. JONGEN, J.-J. RÜCKMANN, O. STEIN, *Generalized semi-infinite optimization: a first order optimality condition and examples,* Mathematical Programming, Vol. 83, pp. 145–158, 1998.

73. H.TH. JONGEN, V. SHIKHMAN, *Generalized semi-infinite programming: the nonsmooth symmetric reduction ansatz,* SIAM Journal on Optimization, Vol. 21, Issue 1, pp. 193–211, 2011.

74. H.TH. JONGEN, V. SHIKHMAN, *General semi-infinite programming: critical point theory,* Optimization, to appear.

75. H.TH. JONGEN, V. SHIKHMAN, *Bilevel optimization: on the structure of the feasible set,* Mathematical Programming, to appear.

76. H.TH. JONGEN, V. SHIKHMAN, S. STEFFENSEN, *Characterization of strong stability for C-stationary points in MPCC,* Mathematical Programming, to appear.

77. H.TH. JONGEN, F. TWILT, G.-W. WEBER, *Semi-infinite optimization: structure and stability of the feasible set,* Journal of Optimization Theory and Applications, Vol. 72, pp. 529–552, 1992.

78. H. TH. JONGEN, G.-W. WEBER, *Nonlinear optimization: characterization of structural stability,* Journal of Global Optimization, Vol. 1, No. 1, pp. 47–64, 1991.

79. J.K. KELLEY, *General Topology,* Van Nostrand-Reinhold, New York, 1969.

80. D. KLATTE, B. KUMMER, *Strong stability in nonlinear programming revisited,* Journal of the Australian Mathematical Society, Series B, Vol. 40, No. 3, pp. 336–352, 1999.

81. D. KLATTE, B. KUMMER, *Nonsmooth Equations in Optimization: Regularity, Calculus, Methods and Applications,* Nonconvex Optimization and Its Applications, Kluwer, Dordrecht, 2002.

82. D. KLATTE, K. TAMMER, *Strong stability of stationary solutions and Karush-Kuhn-Tucker points in nonlinear optimization,* Annals of Operations Research, Vol. 27, No. 1, pp. 285–308, 1990.

83. M. KOJIMA, *Strongly stable stationary solutions in nonlinear programs,* in *Analysis and Computation of Fixed Points,* S.M. Robinson (ed.), Academic Press, New York, pp. 93 – 138, 1980.

84. M. KOJIMA, R. HIRABAYASHI, *Continuous deformation of nonlinear programs,* Mathematical Programming Studies, Vol. 21, pp. 150–198, 1984.

85. B. KUMMER, *An implicit function theorem for $C^{0,1}$-equations and parametric $C^{1,1}$-optimization,* Journal of Mathematical Analysis and Applications, Vol. 158, pp. 35–46, 1991.

86. B. KUMMER, *Lipschitzian inverse functions, directional derivatives and application in $C^{1,1}$ optimization,* Journal of Optimization Theory and Applications, Vol. 70, pp. 559–580, 1991.

87. S. LEYFFER, T.S. MUNSON, *A Globally Convergent Filter Method for MPECs,* Preprint ANL/MCS-P1457-0907, Argonne National Laboratory, Mathematics and Computer Science Division, September 2007.

88. Z.Q. LUO, J.S. PANG, D. RALPH, *Mathematical Programs with Equilibrium Constraints,* Cambridge University Press, Cambridge, 1996.

89. L.A. LYUSTERNIK, *On conditional extrema of functionals,* Mathenaticheskii Sbornik, Vol. 41, pp. 390–401, 1934.

90. J.N. MATHER, *Stability of C^{∞} mappings: V. Transversality,* Advances in Mathematics, Vol. 4, pp. 301–336, 1970.
91. J. MATHER, *Notes on Topological Stability,* Harvard University Press, Cambridge, Massachusetts, 1970.
92. O. MAYER, *Topological classification of continuous selections of five linear functions,* Optimization, to appear.
93. J. MILNOR, *Morse theory,* Annals of Mathematics Studies, Vol. 51, pp. 1–153, 1963.
94. B.S. MORDUKHOVICH, *Variational Analysis and Generalized Differentiation I, II,* Springer-Verlag, Berlin, 2005.
95. R. NARASIMHAN, *Analysis on Real and Complex Manifolds,* North-Holland Publishing Company, Amsterdam, 1968.
96. J.V. OUTRATA, *Optimality conditions for a class of mathematical programs with equilibrium constraints,* Mathematics of Operations Research, Vol. 24, pp. 627–644, 1999.
97. J.V. OUTRATA, M. KOCVARA, J. ZOWE, *Nonsmooth Approach to Optimization Problems with Equilibrium Constraints,* Kluwer Academic Publishers, Dordrecht, 1998.
98. P.M. PARDALOS, T.M. RASSIAS, A.K. AKHTAR (EDS.), *Nonlinear Analysis and Variational Problems,* Optimization and Its Applications, Vol. 35, Springer, New York, 2010.
99. D. RALPH, O. STEIN, *The C-index: a new stability concept for quadratic programs with complementarity constraints,* Mathematics of Operations Research, to appear.
100. S.M. ROBINSON, *Regularity and stability for convex multivalued functions,* Mathematics of Operations Research, Vol. 1, pp. 130–143, 1976.
101. S.M. ROBINSON, *Strongly regular generalized equations,* Mathematics of Operations Research, Vol. 5, No. 1, pp. 43–62, 1980.
102. R.T. ROCKAFELLAR, *Convex Analysis,* Princeton University Press, Princeton, New Jersey, 1970.
103. R.T. ROCKAFELLAR, *Maximal monotone relations and the second derivatives of nonsmooth functions,* Annles de Institute Henri Poincaré Analyse Non Linéaire, Vol. 2, pp. 167–184, 1985.
104. R.T. ROCKAFELLAR, R.J.-B. WETS, *Variational Analysis,* Springer-Verlag, Berlin, 1998.
105. H. SCHEEL, S. SCHOLTES, *Mathematical programs with complementarity constraints: stationarity, optimality, and sensitivity,* Mathematics of Operations Research, Vol. 25, No. 1, pp. 1–22, 2000.
106. S. SCHOLTES, M. STÖHR, *How stringent is the linear independence assumption for mathematical programs with stationarity constraints?* Mathematics of Operations Research, Vol. 26, No. 4, pp. 851–863, 2001.
107. E.H. SPANIER, *Algebraic Topology,* McGraw-Hill Book Company, New York, 1966.
108. S. STEFFENSEN, M. ULBRICH, *A new relaxation scheme for mathematical programs with equilibrium constraints,* SIAM Journal on Optimization, Vol. 20, No. 5, pp. 2504–2539, 2010.
109. O. STEIN, *On level sets of marginal functions,* Optimization, Vol. 48, No. 1, pp. 43–67, 2000.
110. O. STEIN, *Bi-level Strategies in Semi-infinite Programming,* Kluwer, Boston, 2003.
111. S. STERNBERG, *Lectures on Differential Geometry,* Prentice Hall Inc., Englewood Cliffs, New Jersey, 1964.
112. G. STILL, *Generalized semi-infinite programming: numerical aspects,* Optimization, Vol. 49, No. 3, pp. 223–242, 2001.
113. R. STÖCKER, H. ZIESCHANG, *Algebraische Topologie,* B.G. Teubner, Stuttgart, 1988.
114. L. THIBAULT, *Subdifferentials of compactly Lipschitz vector-valued functions,* Annali di matematica Pura ed Applicata, Vol. 4, pp. 151–192, 1980.
115. L. THIBAULT, *On generalized differentials and subdifferentials of Lipschitz vector-valued functions,* Nonlinear Analysis: Theory, Methods & Applications, Vol. 6, pp. 1037–1053, 1982.
116. G.-W. WEBER, *Generalized semi-infinite Optimization and Related Topics,* Heldermann Verlag, Lemgo, 2003.
117. H. WHITNEY, *A function not constant on a connected set of critical points,* Duke Mathematical Journal, Vol. 1, pp. 514–517, 1935.

118. J.J. YE, *Necessary and sufficient optimality conditions for mathematical programs with equi-librium constraints,* Journal of Mathematical Analysis and Applications, Vol. 307, pp. 350–369, 2005.
119. J.J. YE, S.Y. WU, *First order optimality conditions for generalized semi-infinite program-ming problems,* Journal of Optimization Theory and Applications, Vol. 137, No. 2, pp. 419–434, 2008.

Index